THIS BOOK IS DEDICATED TO MY FRIENDS AND COLLEGUES ENGAGED ON BUS WORK AND TO ALL THE BUS WORKERS BOTH PRESENT AND PAST WHO ARE SERVING, OR HAVE SERVED, LONDON TRANSPORT OR ITS PREDECESSORS.

In preparing this manuscript the Author has deliberately avoided reference to any political or trade union controversies arising during the period covered.

A lifetime of London bus work

by

Bob Scanlan

Series Editor - Alan Townsin

The Transport Publishing Company · Glossop · Derbyshire

INTRODUCTION

In a large undertaking like London Transport, amongst the thousands of employees, there are all sorts of personalities. This is to be expected, and in the main highly desirable. In any such large organisation one often hears of the man who came from "outside", i.e. received his training and experience from outside resources. The author has made this effort in an attempt to portray his own experiences as one who obviously gained most of his from "within".

In serving an organisation for half-a-century the author feels there must be something in such a record of experiences from which younger and even future employees in the realm of "bus work" may derive some benefit, if only to feel thankful for the changes that have taken place over the years, and for the conditions of employment which prevail at the present time. So many are apt to take these for granted, and in their ignorance imagine that bus work has more or less always been as it is today.

This manuscript was compiled entirely by the author. Comments, or opinions passed are solely his personal views, and are not intended in any way to cast reflection on London Transport or its predecessors. I am proud when I reflect on the various tasks I have performed in the service of my employers, assisting in some small way to keep the wheels of London Transport moving.

For many years I had cherished the idea that I would like to write and publish a book on my various experiences in bus work. This idea was kindled and brought at least to a start by an event in my work in the Drawing Office which occurred in the latter part of 1962. The then Chief Engineer (Road Services) required a brochure to be composed, describing our new Routemaster vehicles which were then entering service. It was to be written so as to be suitable for giving to special people engaged in transport undertakings from both home and overseas, who were visiting our Head Office at 55 Broadway. I was given the task of assisting a Senior Executive to produce this ready for publication. After I had written many thousands of words, and delved through countless drawings and specifications, and between us we had produced what I think could reasonably be described as a good Ms., I asked whose name would go on the front cover of the finished work. I was told that the Chief Engineer's name would be shown as the Author, as we had worked under his direction, as is general practice in large undertakings such as LTE. This struck me as less than fair, for I had never set eyes on him, nor he on me, whilst I was engaged on the project. So I determined there and then that I would write my own book, and be proud to have it published under my own name, and to have the courage to state my own views on how I saw the many events which occurred during my long career.

I should like to take this opportunity to extend to London Transport and its predecessors sincere thanks as employers over so many years, and cherish the hope that I have given satisfaction always in executing any task I have been called upon to perform. All who are employed by the Executive on bus work, whether Officers, Executives, or even general hands, should be united in a common cause to do our utmost to serve this essential undertaking to the very best of our ability, and by so doing, make London Transport's the finest fleet of public service vehicles in the world.

H.D. (Bob) Scanlan
Isleworth, 1979.

CONTENTS

PHOTOGRAPHIC CREDITS

London Transport	All except those below
AEC	19, 39, 63, 75, 92, 93, 105 (lower).
Bus & Coach	18, 27 (upper right), 31 (upper), 32, 37, 99.
Commercial Motor	59 (bottom right).
Daily Record	60, 61 (right).
C. Degan	30 (upper).
J.G. Gilham	7, 11 (upper), 13 (lower), 14, 17 (lower), 25, 30 (lower), 33, 82, 111 (left column).
Leyland Vehicles	71, 72, 77, 78 (right), 100, 106 (top left).
G.J. Robbins	8 (upper), 17 (upper), 21 (upper), 22.
H.D. Scanlan	10, 41.
Straker Squire Motors	8 (lower).
J.W. Strichlik	35 (left).
Alan Townsin	111 (right), 112, 113, 115.
R.E. Vincent	96.
Unknown	61 (left).

This scene at Mansion House was taken a few months before the last NS buses were withdrawn in 1937 — one is visible facing the camera near the centre of the cross-roads. Four of the then recent STL-type buses queue in the foreground and one of the latest examples, with roof mounted route number box, is coming from right to left. Also evident are LT-class six-wheelers. Note the high proportion of buses to other traffic.

1. Boy Engineer

My entry into 'Bus Work'

I commenced my employment with the London General Omnibus Company Limited on 8th August, 1918, after having been educated at Monega Road Elementary School, East Ham, where I had been a pupil for eight years. A solid general type of education was given at this school, with no frills, but stern discipline, with classes numbering up to fifty in each. I was lucky enough to be in the top class for nearly two years, and was able to receive some commercial training. I remained at school until I had reached my fourteenth birthday, although some scholars left at 13. Boys wishing to leave early had to pass what was called the Labour Examination before being allowed to leave, which in any case could not be before their 13th birthday. Times were hard in those days. The First World War was still raging, and very few lads had the opportunity of having a Secondary School education, because for one thing there were so few schools available. In view of this situation the general pattern that followed was to leave school at 14 years of age, get out to work and earn some money, and then pursue further education at Evening School.

On leaving day school I worked at first for an assurance company in Bishopsgate, London, E.C., as a junior clerk, for a salary of £35 per annum. But after three months I resigned, because over this period I had gradually acquired the desire to become an engineer. Largely due to the war there was a great demand for engineers at this time, and it was considered an excellent trade for a lad leaving school to which to be apprenticed.

So one day I went to Forest Gate garage of the L.G.O.C., which was situated on the west side of Green Street, just over half-way from Barking Road to Ronford Road, and occupying most of the land between Boleyn Road and Upton Park Road. Here I saw the Garage Engineer and explained my mission, cherishing the hope that I would be apprenticed almost immediately. On seeing me he exclaimed: "Very small lad for fourteen", and added "The work here is very hard and extremely dirty". To this I replied that I did not mind hard and dirty work (though inwardly I was not too sure!). I can imagine myself as I was then, a 14-year-old lad, "5ft. nothing" in height, gazing at this gentleman, somewhat in awe, who I was to learn in due course was known to all in the establishment as the "Governor".

This worthy gentleman explained that I would have to start as a "boy", and if I behaved myself and showed promise I would be apprenticed on attaining the age of 16. There were no formalities such as those that are associated with obtaining employment nowadays, merely verbal questions and answers. No birth certificate or references was requested, and there was no medical examination. (This took place about six months later, and is reported on Page 18 of my story.) Only my name, age, and address were requested, and there was nothing to sign. I suppose I must have created a favourable impression, for I was introduced to the dock foreman and told to report to him the following morning at seven o'clock, and so the very next morning entry into "Bus Work" had begun.

My first day

The recollections of my first day of "Bus Work" are still vivid in my memory. I awoke long before 6 a.m., to get up, have breakfast, and be ready to start at 7 a.m. This, to me, only recently having left school, was an unearthly hour, but, as part of the job to become an "engineer", was readily accepted. What was to become my working attire had been laid out the previous evening, and a "brand new" suit of dungarees had been purchased to keep my clothes reasonably clean.

The garage was only about eight minutes' walk from home. I shall never forget setting out, with a man's cap on my head in place of the school cap or Sunday trilby, and wearing heavy boots to which I was totally unaccustomed. However, here I was, strutting along full of enthusiasm to enter the realm of "bus work", somewhat apprehensive, maybe, as to what it would prove to be, but nevertheless determined to do what I was told and to learn all that I possibly could about this work. As I look back now I realise that I was a shy lad, reserved, timid, and rather short in stature for my age. I had lived a very sheltered and happy home life, and had very little experience of the rough and tumble of industrial life of that period.

I left home in plenty of time to allow me to arrive at least ten minutes before starting time, wondering all the while how I should be received and what I should be given to do. I had been told to report to the dock foreman and given his name. On request I was directed to this worthy gentleman, whom I found in his office. This office was, in fact, a timber structure about the size of an average garden shed, equipped with a desk and stool, and adorned with assorted bits and pieces of steel and brass of all shapes and sizes, some strung together and hung on the walls, and others lying loose on the desk and floor. These, I was later to learn, were in fact parts to be used on dock cars, as and when required. For some unknown

(Left) The type of bus that played a major part in the author's career. The LGOC B-type was exclusively operated from all three garages at which he worked. B1619, seen in the original red livery without the white window pillars etc. introduced subsequently, was operated from Forest Gate garage on route 23 and may well have been there when he was. Note the inverted destination board.

(Below) The interior of Forest Gate garage in June 1923. The buses nearest the camera, left and right, are B-types, as exclusively used when the author was there in 1918-20, but many of the remainder are S-types. The dock area is at the far end on the right.

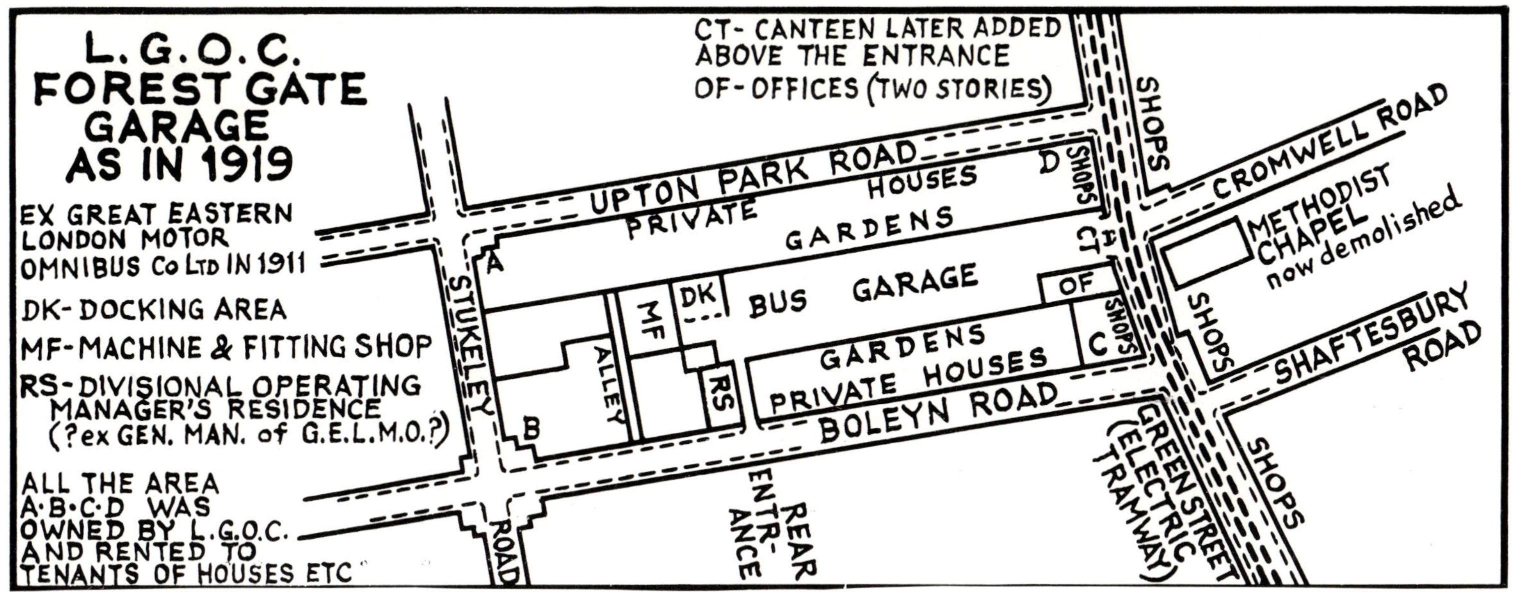

reason all vehicles entering the dock for attention were then, and are still, known as "dock cars".

The foreman, who to me seemed a real old man, though he could not have been over 40, received me in very friendly manner, and promptly took me over to another lad who was probably a little older than I, and to whom he gave instructions to show me "the ropes" as I should be taking over his job. Now this lad was very pleased to see a "new boy" arrive, so as to be in a position to hand over all the dirty, rough, and irksome tasks that he had previously had to do. As I was to learn afterwards, the "last boy" to start had to do all the rough and dirty jobs and to act as tea-boy. This lad said "Follow me", and we threaded our way through parked buses to the wall of what I was later to learn was called the "Dock". Here, amidst an array of jackets, I found an "empty" or unused nail among many which had been driven into the wall to serve as coat hooks. I hung up my jacket and donned overalls, ready and willing to start.

I was then shown "the ropes" and my way round the establishment. I reckon I saw the ropes, string, wire, the lot!, for by the time dinner-time came I had certainly been initiated into "bus work" by the following events:- I had received a hooter-bulb full of water on top of my head, which came from nowhere, crawled under umpteen dirty vehicles, fetched the tea, and been sent to the stores for a "left-handed clout" and a rubber hammer. I had previously been taught to respect my elders and persons in authority with plentiful use of "Yes, sir" and "No, sir", and had often been reminded to "mind my P's and Q's" when talking to people of importance. It was now all foreign to me, for in this work-place everybody was known either by nickname (which could be abbreviated surname), or by Christian name. There were no "Misters". The foreman was known as "Bert", and the Garage Engineer as "the Governor".

In this era the mid-day break at noon was known as dinner-time throughout all industrial undertakings, at least on this, the Essex county side of London. I shall always remember my first dinner-time while at the garage. Living only a few minutes' walk away I was able to walk home. I was so dirty I had to scrub up at the kitchen sink, and sit down on a newspaper-covered chair seat, with a sheet of white American cloth (called leathercloth nowadays) over the linen cloth at my position at the table.

Back to work at one o'clock, with more crawling under vehicles, fetching tea and running errands within the garage. The staff were very friendly, but the language horrified me, especially the adjectives! I remember the foreman, whom I always addressed as "Sir", saying to me some weeks afterwards, "Son, don't repeat some of the

The imposing entrance of Forest Gate garage, photographed around 1908-10 when it was owned by the Great Eastern London Motor Omnibus Co. Ltd. The building had not been substantially altered by the LGOC when Great Eastern was taken over in 1911, and was still in this form when the author started his career there in 1918. The bus in the foreground is on the Ilford to Shepherds Bush route.

A Straker-Squire three-ton model of the type operated by both Forest Gate and Leyton garages by Great Eastern and, for some years, LGOC though all had gone from those garages before the author's career started.

language and things you hear in this place to your mother". Young lads starting careers in those days were often told by their schoolmasters and elders that to start early in the morning to work never hurt anybody, and that there was no disgrace in working hard on a very dirty task. But the amazing feature was that, invariably, the people who were telling you this were never early-risers themselves, and had never been engaged on hard or dirty work.

At last six o'clock came, and with it time to "knock off" — a typical phrase in this period. Ten hours of work completed! I had walked miles, crawled under countless vehicles, and gathered an abundance of grease, oil and mud on my person for the princely sum of twopence-farthing an hour, so the total amount earned was one shilling and tenpence halfpenny. I had been warned that I would have to take the rough with the smooth in starting to become an engineer, but to me, on this my first day, the predominance seemed to be on the rough.

I went home tired, confused by all I had seen, heard, and had to do in such strange surroundings; the unaccustomed smell of the oil, grease and petrol still lingering in my nostrils, and the steady drone of the old B-type engines still ringing in my ears. However, in spite of all the trials of this day, I remained as keen as mustard in my resolve to become "an engineer on the buses". Little did I know then that my association with buses was to last for over half a century.

L.G.O.C. Forest Gate garage, 1918

Forest Gate garage was originally built in 1906, partly as a horse-bus yard; in fact the rings where the horses were tethered could still be seen on the walls. The premises had been acquired by the L.G.O.C. in May 1911 (subsequently made retrospective to 1st January, 1911 as an agreed accounting date), when they took over the business of the Great Eastern London Motor Omnibus Co. Ltd. In horse-bus days, until 1906, this had been known as the Great Eastern of London Suburban Tramways and Omnibus Co. Ltd., which had been a subsidiary of the old Lea Bridge, Leyton, and Walthamstow Tramways Company of horse-tramway days, and, despite its title, was nothing to do with the Great Eastern Railway.

By 1918 the water tanks recessed into the floor at intervals along the walls, which when installed had served to water the horses, were now used to contain the water required for bus washing. The Dock was separated from the rest of the garage by a timber framing which was boarded up to waist level, and above this was covered with wire-mesh. The only purpose this achieved was to define the Dock area, and to some degree it served to keep out the road staff or un-authorised persons. The Dock contained a bench running the whole length of the outer wall, and in one corner a small timber hut served as an office, complete with telephone for the headquarters of the "running shift".

Six pits were installed. These were roughly lined with timber and concrete, and when not in use were covered with heavy boards, often soaked in oil and grease. No lighting was installed, and hand leads were used, connected by means of sockets hanging in the roof. These sockets were reached from the top deck of a vehicle. Steps were fitted at one end of the pits, but these over the years had become saturated with oil and grease. Frequently after heavy falls of rain or snow the pits had to be baled out before they could be used, as there were no drainage or cleaning facilities installed. The cleaners, where all the mechanical units were cleaned prior to overhaul, consisted of two caustic-soda-solution water tanks, heated by an antiquated upright boiler, and pails of paraffin with "diddler" brushes served to supplement the cleaning arrangements.

At this period there was no central chassis-overhaul factory such as Chiswick or Aldenham, and vehicles to be overhauled were driven from Forest Gate to one of the company's coach factories at either Olaf Street, Hammersmith, or North Road, Islington, where the body was dismounted and subsequently overhauled. The chassis was then driven back to the garage, where it entered the "shop" and was stripped down to the bare chassis frame. Each unit when stripped was overhauled by a fitter who was held responsible for the satis-factory completion of the overhaul of the unit. All fitters were allocated a number, which they stamped on a plate which was riveted to the unit. Great pride was taken in the performance of these parts, and I have often seen a fitter lift the bonnet on a bus to see the engine number plate, to ascertain if it was "one of his".

This work was undertaken in the "shop", which was a more modern extension of the original "yard". The "shop" housed the engine test stand, and ten machines, i.e. four lathes, two drills, two grinding wheels, a shaper, and a machine saw. An overhead shaft, bracketed from the wall and driven by a single motor, supplied the power to drive the machinery via a gantry. The heating of the shop was by a cast-iron cylindrical (tortoise) slow-combustion stove, fired by coke and creating far more smoke and fumes than heat. The blacksmith's shop had once been the living room of a private house, and the one-time scullery was now the coppersmith's shop, while the first floor served as a changing room for the women cleaners.

The office of the "Governor" was a brick building erected in the body of the "yard", with the general office and its staff of one man, one woman, and a lad adjoining. The toilets were very primitive, and washing facilities for staff simply did not exist. Oil and grease were partially removed from the hands by paraffin from a bucket in which small working parts had previously been cleaned. If you worked in the shop you were privileged to wash in a bucket of warm water which had previously been heated, either by blow pipe or by standing on the top of the combustion stove.

The inside staff, as the garage staff were called, worked 54 hours a week. This was from 7 a.m. to 5 p.m. on Monday, 7 a.m. to 6 p.m. on Tuesday, Wednesday, Thursday, and Friday, and 7 a.m. to 12 noon on Saturday. There were of course *NO* paid holidays, but work was suspended on Bank Holidays, when the principle of "no work no pay" operated. Two minutes of starting-time allowance was given, but if you were late in excess of this then half an hour's wages had to be lost. Indeed, to be late was regarded as almost criminal. The time recording clock was positioned just inside the garage entrance, and although many employees clocked on well before the recognised starting time they all remained around the entrance until then, when they all trooped up to the workshop in a body. Coats were then removed, and hung on nails around the walls, as no cloakrooms, lockers, or even coat hooks were provided. Overalls were then donned, and work was resumed.

The composite clock cards were a single sheet folded in half. The outer front side was designed to include provision for time recording for a week, and the inside for a record of hours worked per job under various job numbers. On the back of the card, the rate per hour, stoppages, and gross and net wages were recorded. Payment of wages was on Fridays at noon. Cards were distributed previously by the foreman, and, when signed, were presented to the garage cashier, who paid out from a bulk supply of cash.

All disputes and grievances were taken direct to the "Governor", who engaged or discharged men more or less on the spot. He was, and certainly had to be, a real engineer, and a very practical one too. The general foreman of the garage could easily be distinguished as the gentleman wearing a white overall coat and a bowler hat, whereas the shop and dock foremen wore brown coats. Smoking was not allowed anywhere on the premises, and large white notices with red lettering were displayed in prominent positions warning employees of instant dismissal if they were caught disregarding the regulation.

On being engaged, staff were at first issued with a duty pass, entitling them to free omnibus travel to and from their homes and place of employment. The pass would not be issued until a photo-graph of the holder was stuck on to it in the place provided. It was on account of these photographs that the pass was known by the staff as a "sticky", a name which persisted right up to the present "all-routes" travel pass, which was renewed on the 1st of January in each subsequent year, with a different colour for each year.

There was no provision of a canteen for the garage staff, but catering facilities were available in a shop situated on the opposite side of the road, which was owned and controlled by the company, and known as the messroom. No official tea breaks were permitted during working hours, but tea, cakes, and sandwiches could be purchased by ordering from a lad who was allowed to proceed to the messroom. This was me! Each man had his own tea can, usually a beer can obtained from the neighbouring White Hart public-house, which the lad carried on a notched pole. The refreshment had to be taken while still working.

Once a year, on a Saturday during the summer, a staff "beano" was organised, and sponsored by the "Governor", who would grace the venture with his presence. Those participating had the morning off from work, but without pay. Two or three omnibuses were hired from the company (I believe special terms were arranged), and the party proceeded to Brighton, with an occasional stop for liquid refreshment, at almost a flat-out speed of about 20 miles an hour, with the old "B"-type omnibuses almost gasping when gaining the summit of some of the Surrey and Sussex hills. On arrival at the destination the dinner was partaken of as a party, and after a few speeches and drinks we all dispersed, until departure from the sea-front at about eight o'clock in the evening. With more stops en route homeward bound, the party usually arrived back at "The Gate" round about midnight. This memorable day out was for most of the staff, alas, the only voluntary time off they were able to have throughout the whole year.

The staff of Forest Gate garage photographed on their annual outing during the summer of 1919, at a refreshment stop on the Brighton road.

What we did at Forest Gate

Although Forest Gate garage had been acquired from the Great Eastern company in May 1911 there were no ex-Great Eastern buses still there when I arrived in 1918. All the Great Eastern Arrol-Johnstons were scrapped within only a few years, and although the Straker-Squires did mostly survive until 1919 they were transferred away from Forest Gate to other garages in, I think, May 1913. In my time the fleet was exclusively of the L.G.O.C. standard "B" type, which was first introduced in October 1910, although Forest Gate did not receive any until (probably) May 1913, at which date I suspect the entire fleet may have been renewed at about the same time. In 1918 Forest Gate housed about 120 of the B-type 34-seat open-top double-deck vehicles, which seem quaint by modern standards and were still reminiscent of the old horse-drawn omnibus, particuarly in the style of bodywork. They were built on a double-flitch-plated ash frame, powered by a 30 horse-power petrol engine, and fitted with a cone clutch, three-speed silent-chain gearbox, worm-driven rear axle, worm-and-nut steering, and solid rubber tyres. We kept our own fleet self-contained at Forest Gate, as it was not the practice in those days to interchange buses to or from other garages even though they were all of the same standard type. No doubt this was because we were responsible for overhauling them as well as operating them, in the days before the Central Overhaul Works. We did have just one vehicle at Forest Gate that was not a B-type. This was the garage lorry, and was on an L.G.O.C. X-type chassis of 1919. Although most of the X-type survived as buses until 1920 a few of them were converted earlier to lorries, and ours was evidently one of these.

Vehicles operating from Forest Gate garage carried the running letter "G", and were known as coming from "The Gate". We worked on five main trunk routes. These were Route 15 (Ladbroke Grove — Paddington — Oxford Circus — Charing Cross — Fleet Street — Bank — Aldgate — Canning Town — East Ham, White Horse), Route 23 (Marylebone Station — Oxford Circus — Holborn — Cheapside — Aldgate — Canning Town — East Ham — Barking Broadway), Route 25 (Victoria Station — Bond Street — Oxford Circus — Holborn — Cheapside — Aldgate — Mile End — Stratford — Forest Gate — Ilford — Seven Kings Garage, extended on Sundays to Chadwell Heath), Route 40 (Elephant & Castle — London Bridge — Aldgate — Canning Town — Upton Park — Forest Gate — Wanstead, The George), and Route 101 (North Woolwich — East Ham — Manor Park — Wanstead, The George). It is noteworthy that all five of these routes are still substantially the same today, more than 60 years later. Five of the ten ends have been slightly lengthened or shortened, but the main part of each route is still exactly the same.

The continuity of London bus routes is underlined by this photograph taken in 1978 showing Daimler Fleetline DM 1165 on route 101 crossing the opening bridge over the entrance lock to King George V dock, a point where the author refers to delays to the 101 service 60 years earlier.

The 15 was our most important route, needing nearly 40 buses. We were the only garage on the 15 and the 101, but Seven Kings as well as Forest Gate worked on the 25. I cannot be sure about the other two, but it is possible Athol Street (Poplar) garage may have helped out on the 23 and 40. The Gate ran on routes 25 and 40 for half a century, but our workings on the other three were taken over at a fairly early date by Upton Park garage, which the London Road Car Co. Ltd. had built in 1907. The 101 service was frequently held up very badly by the opening bridges over the entrances to the Royal Albert Dock and King George V Dock. Along the full length of Green Street, past our garage, we had severe competition from a West Ham Corporation electric tramcar service from Canning Town to Leyton, and these tramcars had 56 or 60 seats as compared with only 34 on our buses, and covered tops instead of open.

The mechanical work at the garage was under the control of the District Engineer, who was based at Leyton Garage, itself also ex-Great Eastern. His visits were very rare, and he invariably came in a motor-car, which in those days made him somebody of note. Nobody working in Forest Gate garage took much notice of him, although it was clear to all of us when he came, for the presence of his car made that point quite clear. Although this gentleman was a superior to our "Governor" this fact just did not mean a thing to the rank-and-file of the personnel. The office of the Company's Chief Engineer was situated at No. 9 Grosvenor Road, Pimlico, the head office of the L.G.O.C., this being in the days before the construction of 55 Broadway or even the removal to the adjacent Electric Railway House. Should any employee be summoned to attend there, such as apprentices attending for the purpose of signing indentures, it was said that they were seeing "the Heads up the road".

Accidents occurring in our garage were dealt with by the store-keeper, who had the necessary equipment in a corner of the stores. While not wishing to cast any reflection on this gentleman's first-aid efficiency it must be admitted that he was one of the very few employees in the workshop area who was in any way clean enough to handle any casualties. The first consideration was to have reasonably clean hands to deal with patients, and first-aid qualifications, while recognised as being desirable, were of secondary importance. Thank heaven the Cottage Hospital was not far away!

The starting age for employees in the workshops was 14 years, and, by contrast with present-day regulations, lads of this slender age were allowed to use tools, and in fact to undertake any task that the foreman in his wisdom had assessed they were capable of performing. To be quite fair, the lads on first starting were given jobs such as taking dock-car unit numbers, and general fetch-and-carry jobs including tea-boy. The first real task they undertook on a vehicle was usually acting as "head dock-car greaser". This entailed filling up all Stauffer lubricators in such inaccessible places as the rear-axle brake cams with thick evil-smelling yellow fat. After a period of doing such jobs as these, the lads usually progressed to such tasks as valve grinding, dismantling and cleaning silencers,

Forest Gate garage took on this appearance in 1926 and did not alter until it was closed in 1960. This photograph was taken in 1936, three years after the formation of London Transport — note the tramlines in the foreground, a source of competition in LGOC days.

This photograph of B2026, built in 1912, shows the seven-slat lifeguard between front and rear wheels adopted that year which made it necessary for the author to crawl under buses when greasing during his days as a 'boy' at Forest Gate garage. The experimental front lifeguards also shown in this 1916 photograph were not standardised, but the curved vertical polished slats in front of the radiator, curiously reminiscent of American cars of a much later period, formed one version of a feature that was to remain for some years. The bus was allocated to Willesden garage, where the author may well have come into contact with it in 1921.

changing petrol tanks, and then they came on to be a fitter's mate. The apprentices had priority over other lads in claiming to work in the shop.

All buses at this period of time had a very high chassis frame, with plenty of space underneath, which resulted in the rear wheels being very exposed, highly dangerous to any pedestrians who might accidentally fall in front of them. Hence in order to prevent such accidents the space between the front wheel and the rear wheel on each side of the bus was blocked by an enormous lifeguard comprising seven horizontal slats fixed to five vertical struts. But from the point of view of us boys working in the garage these lifeguards were a nuisance, they prevented us from having easy access to the mechanical units underneath. Whilst working in the dock as head greaser and odd-job boy I had on very many occasions to crawl underneath the rear axle in order to reach a grease Stauffer up at the front end of the bus, and I remember thinking then what a problem these lifeguards were, and how I wished that the fitters could remove them for me from one side. But this of course could not really be expected, as time and money were involved, for they were secured to the chassis frame by ten 7/16-inch diameter bolts.

However, one side did have to be removed if a gearbox had to be changed. A tripod hoist would be fitted over the opened trapdoor in the lower-saloon floor above the box, and when the front and rear couplings together with the fixing bolts had been removed the unit would be lowered to the ground and slid out to one side where the lifeguard had been removed. But silencer baffle joints had to be changed without removing the lifeguards. The inspection pits in the garage floors at this time were quite short, probably less than half the length of a B-type vehicle, as compared with the present-day inspection pits which usually extend the entire length or more of the longest vehicle they are intended to accommodate.

The issue of company tools in those days was very meagre, and generally speaking all employees had to supply their own. Tools were very costly, and unlike the present times were usually of a high standard, but there was little variety, and very limited competition existed between manufacturers of such tools. Bolts and metal screws were predominantly of Whitworth thread, and all dimensions were to English measurement. Paper-work to the man in the workshop, with the exception of his very important pay card, was non-existent. Rates of pay were, in general, rather more favourable than those paid elsewhere in the locality. The amount paid was for the actual hours worked, and there was no bonus or sick pay. I personally was engaged at the rate of twopence-farthing an hour under the classification of "boy". If by 16 you were not an indentured apprentice you became a "youth", and over 18 you were an "improver".

The greater part of the staff had been recruited locally, and in the main they were of typical East of London type, genuine, forthright, and no nonsense, enlivened with the Cockney spirit and humour. Technical knowledge was very limited, but practical ability and craftsmanship were unsurpassed in "bus work". There were no sporting activities in direct connection with the garage, but practically all the staff were keen supporters of "The Hammers", the West Ham United Football Club, whose achievements were closely followed, and which formed the topic of many an argument.

During the winter, buses would often come into the "Dock" straight from service, covered with snow and filth from the road, and one of the other boys or myself would be given a task on the vehicle straight away, such as perhaps to change the silencer or grease the brake cams. This entailed crawling underneath the vehicle, and one can easily imagine the filthy condition of our clothes! We would endeavour to warm our frozen hands on the engine radiator if it was still warm. It is interesting to note here that everybody working in the garage without exception wore a cap. This was most essential in order to afford at least some protection against contact with the filth and grease. Some of the caps almost shone with the grease that had accumulated on them.

It was a great thrill for one of us boys to be told by the foreman to go out and meet a vehicle on service and change the engine fan belt (which was flat belting and constantly breaking), or perhaps to change an engine sparking plug, if the driver had phoned through that a change was necessary. We boys could always rely on the driver giving us assistance should we need it on any of these excursions. It was a red-letter day to us to be instructed to accompany the fitter-

Bus washing machine, 1926 style. This was the method used to wash an open-top NS-type at Forest Gate a few years after the author worked there.

The interior of the Forest Gate garage in January 1960, three months before it was closed and the services transferred elsewhere. The fourteen RT-type buses visible have now also passed into history.

tester on a vehicle to be taken out on a road test. This entailed principally to stand on the platform and act as conductor to keep off any would-be passengers. The brake linings would often be "burnt in" on the quiet country roads out beyond Ilford, and various adjustments made, and the performance of the vehicle in the hands of the tester was followed with a great sense of pride.

Summarised thus briefly are the conditions prevailing at Forest Gate garage at this time. Hours were long, and work was arduous and very dirty. The comfort of the employee seemed not to be of any consequence, but in spite of this the general atmosphere was very cordial. The "Governor" commanded respect, partly through the power and authority he controlled, certainly unlike present-day conditions. A personal interest in the operation and maintenance of the vehicles was shared by all. Staff were very conscious of the fact that their job was to keep the buses running, and so to bring in the revenue to the mutual benefit of both the company and the staff. A true family spirit prevailed, and a far keener interest in one's own particular task was manifest throughout the whole personnel of the establishment than is the case today. Such phrases as "our garage" or "one of our buses" were often heard and were meant in a wider sense.

Fifteen years after closure as a bus garage, the Forest Gate premises in 1975 showed little change from their 1926 design, though at least one more name had come and gone over the doors. The days of pride in 'our buses' passing beneath had long gone.

Mishap with a magneto

I had been working at Forest Gate garage for about six or seven weeks, and had not as yet progressed away from the grease can and the tasks of tea-boy and errand-boy. Then one day I experienced an encounter which at the time had me deeply concerned and worried. I was summoned by the foreman, and told to stop whatever else I was doing because he now had an urgent job for me. In those early days at our garage it was customary for all the fitters and unit adjusters and most of the lads, to wear the so-called "bib and brace" overall trousers whilst working, and with a dungaree jacket. I realised from the foreman that he wanted me quickly, and hence in my haste to respond I removed my overall jacket but kept my overall trousers on, and put on my coat and reported to him.

He produced a magneto, as was fitted to all our buses, and told me to take it "straight away" to Old Kent Road bus garage, as they were urgently waiting for it there, and had a bus off the road waiting for it to be fitted. The No. 40 bus service ran down Green Street past our garage, with a stop in both directions right outside. I crossed the road and boarded a southbound 40 going to the Elephant and Castle, a famous public house in south London where it terminated. The magneto was a fair weight for a lad of my size to hold, but I clutched it tightly. The bus was crowded, and I had to stand in the gangway and hold the weight and keep my feet as well. All went well for some considerable way, until the bus lurched, and so did I, and I fell right onto the lap of a portly lady, still holding the magneto. Was I embarrassed, indeed. This was bad enough in itself, but to my dismay I then realised that I was still wearing my greasy bib-and-brace overall trousers.

The lady called the conductor, demanded my name and address, and complained that my dirty overall had soiled her dress. She said a lad of my age ought not to be holding such a "heavy thing", and stated in none too friendly terms that she would sue the Company and me for compensation. The conductor said, "Give her your name and address, Son, I will have to make a report". Was I concerned! I thought, now I am in real trouble. I had visions of being hauled before "our Governor" (the Garage Engineer), and sacked on the spot, which was quite feasible at that time, and thereby all my ambitions of becoming an indentured apprentice of the Company would be gone for good. I continued my journey, and had to change at the Elephant onto a 53 bus which took me down the Old Kent Road. I found the garage, off a little way down a side street, and delivered the magneto to the foreman there, and brought back with me a docket to acknowledge that he had received it.

All the way on the journey back I was worrying, whatever would happen now? When I got there, I told my own foreman exactly what had transpired, and, good fellow that he was, he allayed my fears and told me in a kindly tone not to worry. He said it could not be helped, and he was sure our Company could handle the situation and not blame me. I felt relieved after telling him, and in actual fact I never heard another word about the occurrence at any time afterwards.

From time to time around this period I was sent on various other errands to garages, either to collect some item of equipment or else to deliver some small spare parts that did not justify sending a lorry. On one such occasion I was sent to Seven Kings garage, in Seven Kings High Road, which at that time was the terminus of the 25 bus route, to deliver a small package. I duly handed it over to the person I was instructed to, whose office I discovered was well inside the confines of the garage. When leaving the building I was surprised when the warden at the entrance called me over to him. I thought, what have I done now? He then said to me, "Son, you don't want to go back with this", and promptly removed a long length of green felt which "somebody" had mischievously hooked onto the back of my coat collar. I felt very foolish when I became aware that I had walked all round the garage so adorned, and I was very thankful that the kindly warden had drawn my attention to it.

These various excursions gave me a welcome break, to enjoy a bus ride and to have an opportunity of going to other garages of our Company. For the most part they were quite uneventful. Sometimes I was sent to the A.E.C. factory and stores, which in those days were situated at Blackhorse Road, Walthamstow, to collect some small items which were urgently required at our own garage and which I could reasonably be expected to carry. I was always received here with every courtesy, and never experienced any practical jokes whilst visiting this establishment. The reader must remember that the First World War had now been running a very long time, more than four years, and spare parts required for running the Company's fleet were now in very short supply. Pre-war stocks of spares had been used up, and new stocks could not be built up in the way our Chief Engineer would have preferred, and so we had to more or less live a hand-to-mouth existence. Output from the Walthamstow factory was enormously greater than it had been a few years previously, but almost all the parts made there now were needed to keep vast fleets of Army lorries running in France and Belgium.

Armistice Day and the lamp cowls

A date now very famous in British history is the 11th of November, 1918, the day on which Peace was declared at long last after those terrible 4¼ years of the 1914-18 Great War. For us at Forest Gate bus garage the day started in the normal way and to the usual routine and pattern. It was a Monday morning, but while we all started work we all were aware that "something" was in the air on the battle front in the War. I realised that all the men were engrossed in talk and discussion for far longer than they would have dared in the ordinary way. As a young lad of only fourteen years of age, who had been working for the Company for only just three months, I had nevertheless seen some of the tragedies of the War. I had witnessed an enemy aeroplane, and also a Zeppelin airship, falling down in flames after being hit by our gunfire, not very far away from where we then lived. I had also experienced the bombing and the anti-aircraft gunfire during many air raids, for we lived on the direct route from the east coast to London. So I was well aware that the sooner the war was over the better it would be for all of us.

However, on this eventful Monday morning, we knew nothing officially as yet, so it was Carry On as normal. Tea must still be fetched for the workmen as usual, and there was no reason or excuse to stop this. I still had to attend to the greasing of the vehicles waiting for me in the Dock area, and also I was still conscious, as usual, that I must be available all the time, ready and willing to be summoned to execute some special service or other, which was not always to me a very agreeable one. Reflecting backwards nowadays I realize how dedicated I was at that time in my struggle to become an engineer on "bus work". I think it is now a great tragedy in these modern days that interest and dedication to one's job or occupation are so rare, especially since lads now leave school so much later and should therefore be "supposed" to be better equipped to face the problems of earning a living.

But I have strayed. We must return to that memorable Monday. The atmosphere was tense and we knew something was going to happen, even though we did not know what. Then eleven o'clock came. It was the eleventh hour of the eleventh day of the eleventh month of the year 1918. Suddenly the rockets sounded all around us everywhere. All the vessels in the Royal Albert and the Royal Victoria Docks at North Woolwich only just over two miles away sounded their sirens long and loud. What a lot of noise! There was no radio in those days, and very few telephones, but the news got round very quickly nevertheless.

In the garage we all continued to work, or attempted to work, for one more hour, until our normal dinner time at noon. I went home to dinner as usual, and when we all returned at one o'clock, the men, through their elected shop steward, saw the "Governor" and requested permission to have the rest of the day off. This was granted, to all except those on urgent jobs. But in my case I was not too keen to go home so early and thereby lose money, and hence I was somewhat relieved when the foreman collected some of us young lads together and said the "Governor" had a special job for us to do. He then explained that we would be much later home that evening than usual, but what actual time he could not yet say. He was a very considerate fellow, and he allowed those of us who lived reasonably near the garage to pop home quickly and inform our parents, if we so desired. I remember how I went home and explained to Mother that I had been assigned to go on a "special job", and would be much later home than usual, but she was not to worry. I was thrilled, and relished the prospect of a new experience in my career.

When we had all returned to the garage, another lad and myself were told to go into one of our buses which had in our absence been loaded with a lot of big bulky cartons. The driver was to be one of our unit adjusters with whom we were both well acquainted. We boarded the vehicle, and were driven to the terminus of the No. 15 bus route at the "White Horse" public house, in High Street South, East Ham, at the corner of Central Park. On the journey there I sat on the exalted seat next to the driver, which was always a thrill for us young lads. In those days, of course, the B-type (and most other buses) had the driver behind the engine instead of alongside it as on later types, and hence there was a seat running the full width of the bus, in front of the bulkhead and in front of the main saloon. The driver sat on the right-hand end, whilst the left-hand end and the centre were not normally used except as-and-when required by Company servants when travelling on duty, especially inspectors. Indeed it was usually called the "inspector's seat", and the space underneath was occupied by the petrol tank. On our trip to the 15 terminus the other lad stood on the rear platform, no doubt equally thrilled to imagine that he was acting as a conductor. His task was to keep off all would-be passengers, saying "Private" if challenged.

On arrival at the White Horse we discovered that all the cartons that had been loaded into our bus contained the original glass lampshades for the interior lamps of a large number of buses. These had all been carefully removed at some period during the War, and had been stored in the garage "for the duration", and at the same time replaced by "black-out" cowls which shielded most of the light from the exterior of the vehicles so that they could not be seen from above by enemy aircraft. The Company's Chief Engineer at Grosvenor Road, Pimlico, had issued instructions that the original shades were to be re-fitted immediately the war had ceased. This quick action was typical of our Company in those days, so unlike any action taken in modern times which is usually preceded by several meetings, plenty of talk, and heaps of paper-work before it is even started.

We two lads were instructed by our driver, and under his guidance we invaded each No. 15 bus as it arrived at the terminus. We had to remove the sheet-metal black-out cowls from every lamp, and re-fit the original glass shades. We soon found it was quite easy to do this, and I recall kneeling on the seat cushions so as to reach the lamps. The drivers and conductors of each bus as they arrived greeted us with good humour, and saw we were supplied with any refreshment that we cared for. Every bus driver in those days carried a blue or white enamel canister full of tea, for drinking at the terminus or elsewhere, and at certain strategically-placed cafes along the line of route the conductor would quickly pop out and purchase a refill. The 15 was a long route, so the crews had a long layover at East Ham White Horse, about 10 or 12 minutes as near as I can remember, which gave us plenty of time to deal with all the lamps, although we really had only six minutes for each bus because the service operated every six minutes and there were always at least two buses on the stand at any one moment.

As the afternoon advanced each group of passengers alighting from the bus as it arrived at our terminus became more and more merry and excited than the last, no doubt because of the stronger liquid refreshment that they had consumed. Being November it got dark very early, and bonfires were soon being lit in the surrounding streets, and we could see the reflections in the sky of other bonfires further away. People in the streets everywhere were singing and shouting, whilst others were removing their own black-outs from the windows of their own houses and shops.

I cannot now remember how long it took us to complete our assigned job on the buses, nor at what time we finished, but I do know that when we eventually arrived back at our garage I was flushed with success and pride at having been associated with the completed operation. We had to deal with every bus on the 15 service, and stay at the White Horse terminus (on an island in the widened entrance to a side turning opposite the pub) until the first bus we had treated had been to the far end of the route and got back to us again. The other end of the 15 at that time was Ladbroke Grove, "The Eagle", the whole route being the same as it still is today sixty years later. Our garage (Forest Gate) was fairly near the East Ham terminus (nowadays it has been replaced by Upton Park). At various subsequent periods the 15 has also been partly worked from the far end, by Middle Row or Willesden garages, but from memory I feel pretty sure that in 1918 our garage had the entire service and no buses came from any other garage. Certainly we changed the lamp shades on every bus as it arrived, and there was no question of "this is one of ours so we do it, that is one of theirs so we don't do it". And certainly there was nobody at Ladbroke Grove changing lamp cowls, and I don't think we at our end changed any on behalf of X or AC. According to the L.G.O.C. bus map at that time the 15 service was advertised to run every six minutes, with a journey time of 104 minutes (and a throughout fare of only ninepence!). Allowing for ten minutes layover at each end I calcu-

Buses on route 15 still terminate at East Ham (White Horse).
The inn sign is just visible on the right of this photograph
taken in July 1978, showing RML 2333 and 2561 on the route,
together with DMS 1452 on route 147.

late that a round trip would take 228 minutes, so I suppose we two lads must have encountered a total of 38 buses and that the whole task must have taken us not quite four hours.

At any rate, after all this I was credited with some overtime payment. I am sure our Foreman was as generous as he dared be, which as far as I can remember was a lot of money to me, in those days. I learnt on the morrow that two other lads from our garage had been engaged on the same task as us at the terminus of Route 40 outside "The George" public house at Wanstead. The 40 was also a Forest Gate route, running through to the Elephant & Castle, but I think would have needed only about 28 buses. In fact the entire fleet of the company was similarly converted on that one day, Armistice Day, and was similarly carried out by lads from all the other garages stationed at suitable terminal points. Little did I know at that time that another war would be starting not quite 21 years later, and that I was destined to be again involved with black-out cowls, this time directly responsible for manufacturing enough cowls to fit the whole of a much larger fleet of buses, as I shall relate later in this book.

2. Apprenticeship

My medical examination

After all these years I still cannot refrain from smiling when I recall my medical examination. After working for about six months at Forest Gate Garage (I was by now working on dock cars), the dock foreman came up to me and said, "Son, you are wanted in the 'Governor's office". Now at this time I was occasionally sent for by the "Governor" to go to the messroom across the road to get tea and biscuits for him, with the message:- "Tea for one, or two, which ever it was, for the "Governor". This involved the steward getting out the best china, kept for such occasions, together with the silver tray and tea pot.

However, on this occasion the chief had another gentleman with him, and he turned to me and said "Son, this gentleman is a doctor, please reply to the questions he asks you". This worthy sir, immediately asked me "How do you feel?" On hearing my favourable reply, he promptly asked to look at my tongue. After a very brief inspection he continued to ask me if I had lost any time from work. On learning from me that I had not, he straight away informed me, "That will be all". The whole examination probably did not last two or three minutes, and so ended my medical examination, clad in grease and oil-covered overalls, cap in hand, straight from the workshop to the Engineer's office.

The doctor's report was evidently favourable, and I presume he considered that I was fit and tough enough to stand the hard and rigorous conditions imposed on the staff, for I received no further call for examination. I must admit on first starting at the garage on one or two occasions I suffered from sickness, due to being unaccustomed to the evil smell of the oil and grease, but I soon got toughened up, and never lost any time, for to lose time was to lose money.

My indenture is signed

One day in September 1920, when I had been working at Forest Gate Garage for just over two years, the Garage Engineer sent for me, and explained that arrangements had now been made for my father and I to go to Electric Railway House, Westminster, to sign the necessary indenture for me to be apprenticed to the Company. The Engineer who had started me on my career in bus work had by this time left the service of the company, and as soon as another Engineer had been appointed I had very quickly reminded him that

Mr George J. Shave, Chief Engineer of the London General Omnibus Co. Ltd.

I had been promised an apprenticeship by his predecessor. He readily agreed to honour this promise when he saw how eager I was to be bound by mutual contract with the company and to learn my trade.

Hence on 25th September, 1920, I left work at dinner time (12 noon), and went home, and after a meal and a scrub up I donned my best suit, and, accompanied by my father, we proceeded by train to St. James Park Station. On arrival at the near-by Electric Railway House we were escorted to the office of the Chief Engineer, Mr. George J. Shave. This gentleman greeted us most cordially, and began to ask me how I liked my present job and what I had been doing since I had been in the company's employ. He was most interested in what I had to say, and encouraged me to continue to work hard and to continue my studies at evening school. He then showed my father the Indenture, which had been lying on his desk, and requested him to read it. After doing so my father, to my dismay, explained he was not satisfied with it, as my name was wrongly spelt. The Chief saw his point, and immediately gave instructions to have another one prepared right away. This was duly signed by us all, and my father then paid the necessary £5 deposit as a guarantee for the satisfactory completion of the apprenticeship in accordance with the terms of the Indenture.

This amount of money in those days seemed to me a really princely sum. A receipt for the money was given and received, and we all shook hands cordially, and my father and I than left for

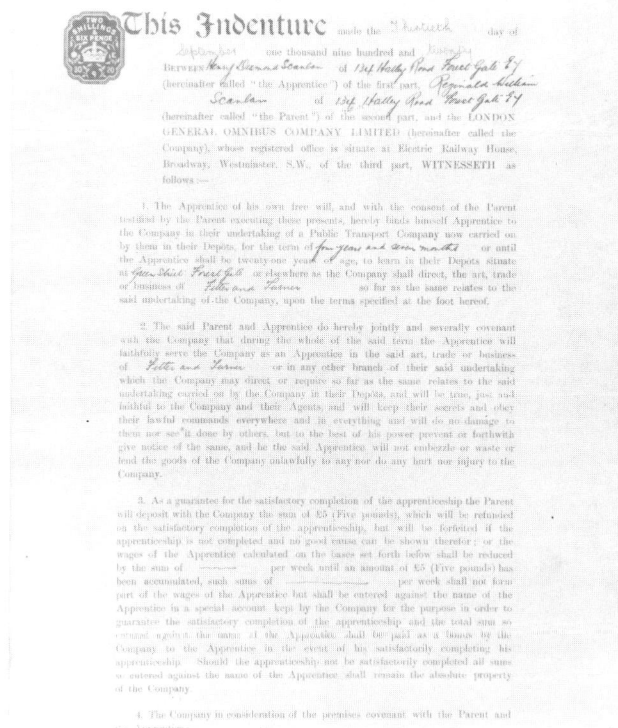

Indenture). I had now spent two years at Forest Gate, and the move became necessary because it was found that there were eight apprentices at Forest Gate, and only one at Leyton. Hence as I was the last to be indentured I was the one to be transferred. I already had a fair knowledge of the locality, because I had lived as a youngster in no fewer than four different houses in the area and had been a pupil at both Farmer Road and Capworth Street Schools.

Leyton Garage was situated where Leyton High Road joined the High Street, and it occupied most of the site between Canterbury Road and Cheltenham Road. In front of the main entrance was a triangular shaped green, and shrubbery. By reason of this the garage was often referred to as "The Green". The running letter for vehicles

Almost indistinguishable from that of a horse-bus, the rear view of a B-type bus standing inside Leyton garage during the time the author worked there.

home. I was genuinely delighted, and had at last achieved my ambition to be apprenticed to become a real engineer. I had lost half a day's wages through going to see the Chief, but I thought that now when I go to work tomorrow I am no longer graded as just a "boy", but as an indentured apprentice of the Company. I felt very proud of myself, and I remember I made up my mind there and then that I was going to see that nobody was going to stop or hinder me in my efforts to learn the art or trade I had chosen.

L.G.O.C. Leyton garage, 1920

I was transferred to Leyton Garage immediately afterwards, at the end of September 1920, having now achieved my boyhood ambition of becoming an indentured apprentice to the company in "the art, trade, or business of fitter and turner" (an extract from the

operating from the premises was "T", and we worked on Route 35 (Clapham Common – Brixton – Camberwell – Elephant – London Bridge – Shoreditch – Dalston – Hackney – Lea Bridge Road – Leyton – Walthamstow) and Route 38 (Victoria – Piccadilly – Bloomsbury – Islington – Dalston – Hackney – Lea Bridge Road – Leyton – Walthamstow – Woodford). On Sundays the 35 was extended to Chingford and the 38 to Epping Forest. As also with our five Forest Gate routes, the 35 and 38 are still substantially the same today in 1975 as they were in 1920, or indeed as when they were first started in 1910 and 1912 respectively.

The garage was originally built by the Great Eastern London Motor Omnibus Co. Ltd. in 1906, to keep its new motor-buses separate from its large horse-bus fleet, at a time when big expansion was starting, but, like Forest Gate, it was taken over in May 1911

Leyton garage of the LGOC. This photograph was taken in June 1926, some of the drivers wearing their summer-issue white coats.

This photograph of the interior of Leyton garage was taken during the 1920-21 period when the author was there. B702, in the early livery in the foreground, had just been given a wash by traditional bucket-and-sponge methods, the other B-types being in the later livery with white window frames.

(back-dated to 1st January 1911 for accountancy purposes), by the L.G.O.C. The General very soon closed Leyton garage, and transferred the buses (which had been mostly Arrol-Johnstons and Straker-Squires) temporarily to its own Clay Hall garage at Old Ford. But Leyton garage was reopened in June 1912 after being completely rebuilt and enlarged. It had been designed as a garage to house motor-buses, but unfortunately the planners were still thinking in terms of horse-bus yards, and the structure and general layout of the premises at Leyton followed the basic style of Forest Gate, which had been built six years earlier. No changes had been adopted which in any way improved the conditions or comfort of the employees. There were still no washing facilities, and the same primitive procedure of heating a pail of water on the top of the slow combustion stove in the shop was followed.

By this time the adoption of the 47-hour working week had been accepted by the company for the engineering staff, and the revised hours were from 7.30 a.m. to 5 p.m. Mondays to Fridays, with an hour's break for lunch from noon to 1 o'clock, and 7.30 a.m. till noon on Saturdays. With these new reduced hours the taking of refreshment in any form during working time was strictly forbidden. There were no canteen facilities available, but a privately-

(Above right) B1609 operating from Leyton garage on route 35, another route that still runs today, though only from Clapham Common to Hackney, rather than to Leyton and Walthamstow. The Leyton fleet was entirely of B-type buses in the author's days there.

(Right)
Another interior view of Leyton garage, taken about sixteen years later, in 1936. The style of the doorways and windows clearly belong to pre-World War I days, but a clean air of good housekeeping is evident.

The K-type bus introduced by the LGOC in 1920. This one was allocated to Forest Gate as signified by the G running number plate, but not until after the author had left there. It was photographed on route 15 (East Ham to Ladbroke Grove) around 1923/4. Apart from introducing the forward-control half-cab layout, it incorporated straight-sided side panelling instead of the rocker-panel style of the B.

owned coffee shop on the opposite side of the road catered for the needs of the staff as far as possible.

At around this time, 1920, the task of re-equipping London's bus fleet was embarked upon by the company, which decided to replace a substantial number of the now ageing B-type vehicles by introducing a new double-deck open-top vehicle mounted on a lightweight chassis and powered by a 28 h.p. petrol engine. This chassis was the first post-war type produced by the Associated Equipment Company, Limited, at Walthamstow, and it was designated the "K" type. The engine with its auxiliaries was mounted slightly to the near-side of the vehicle centre-line enabling the driver to sit beside it, instead of behind it as on all previous types except the Wolseleys and Clarksons (which never ran at Forest Gate or Leyton). This layout provided more body space, and the straight-sided box-like body seated 12 more passengers than the B-type.

As on subsequent types also, the driver's seat, which was a shaped piece of plywood (he had to supply his own cushion), was incorporated on the top of a box-like timber container which was situated in front of the bulkhead across the entire width of the body, and which housed the petrol tank of 26 gallons capacity. A waist-high metal scuttle with a canvas apron above it was the only protection against the weather afforded to the driver, apart from the cab roof above his head. Solid tyres were still fitted. These vehicles were first operated from Seven Kings garage (on 26th August, 1919), and as yet had not arrived at Leyton.

Also at about this time the idea of a central overhaul works, introducing the latest mass-production methods, and applying them to bus overhaul, had been born, this being an entirely new idea which had not previously been attempted. Work had started already on a 31-acre site at Chiswick, and in preparation for this a gradual splitting up of the operations necessary for the overhauls of units was taking place at the various garage workshops. At Leyton garage this had already been applied, but only to the engine. I myself, as a first-year apprentice at the rate of three-pence halfpenny an hour, with the assistance of another youth (not an apprentice), was completely overhauling all rear axles and brake gear. We also executed all repair work on dock cars involving brake cam and brake levers. All work was inspected by the shop foreman, to whom all queries regarding procedure were addressed.

All the machines in the shop were driven by a single motor through a main shaft to their respective countershafts. Hence, should the motor fail, all the machines were unable to be used. The power

was transmitted by flat leather belting, which when it failed to grip the pulleys was treated with "black jack", an evil-smelling sticky substance which was used to assist the belting to grip. The main shaft and countershafts were lubricated by inverted "bottle and needle type" lubricators, while all machines had to be lubricated by their operator with an oilcan. The machines were very old, but in the very capable hands of various members of the staff they produced some really excellent work.

We had no centre grinders, and all turned work had to be finished on the lathe. The tools for the lathes and shaping machine were made by the garage blacksmith, whose forge and anvil were, in fact, situated in a corner of the machine and fitting shop. The tools were made of carbon steel, as no high-speed steel or carbide tips were then in use, and they were then ground and finished by the machine operator. The grinding had to be accomplished on an old-type emery wheel, which itself had to be "hand dressed" with wheel-dressers, and the tools were then finished off by hand stoning.

All press work had to be accomplished on a hand press, and all lifting by chain hoists. There was no welding equipment. Fractured crankcases, gear-boxes, engine cases, and differential cases were plated and screwed, and some very ingenious repairs were effected by this method. The scrapping rate of parts was particularly low, and no item was scrapped that could reasonably be repaired. Fractured chassis frames, whether on dock or overhaul vehicles, were plated on both sides, and drilled by hand with ratchet brace and pillar, and bolted by ½" diameter bolts and nuts. There were no portable electric or pneumatic drills available.

To effect a change of front or rear axle on a vehicle, bottle-type hand-operated screw jacks were used, and the structure was supported under the chassis frame on heavy timber trestles. The age-long problem of endeavouring to keep bolts and nuts securely tightened was solved by the bolt heads being drilled and wired, also by the use of spring and fastnut washers, and of course the method which is still frequently used — the castle nut and split pin. Self-locking nuts of several different patent types which are extensively used today were not manufactured until the early thirties. In some instances the ends of bolts were ground or filed flush and then locked by centre-punching the end of the screw thread adjustment to the nut, and peening over the bolts for locking the nuts was also practised.

At this period stocks, dies, and die-nuts were not easy to obtain, so the staff would make die-plates from old flat files to enable them to run-down damaged screw threads. Torque spanners had not yet been introduced, but fitters knew just how much to tighten nuts without breaking the bolt or stripping the thread. Experience had taught them this. No protective anti-corrosion treatment was used on bolts or nuts, and this was not generally adopted on buses until after the Second World War. Chassis frames and cross members were brush painted with quick-drying black paint which served to protect them from rust. There were no self-starters on buses around the 1920 period, and they all had to be swung by hand. It is easy to realise that the condition of the starting handle and the engaging dogs had to be very good, and indeed this was most essential.

It was most noticeable that when a major repair on a vehicle or unit was envisaged the whole resources of the establishment were harnessed in unison to achieve a success of the operation. While admitting that occasionally some tasks were "bodged" or "wangled" to effect a temporary repair, for the most part the standard of work produced was high. It was an easy matter to ascertain who was responsible if a unit failed in service, unlike modern times, when the blame is hard to pin-point and individuals concerned continue to pass the buck. There were very few "spare" buses available, for the maxim was that buses were built to carry passengers and subsequently earn money, not to be "up against the wall".

My career at Leyton garage came to an abrupt end. The working week for employees started as from Wednesday morning and finished Tuesday night. One Tuesday afternoon early in April 1921, just before four o'clock, the shop foreman came to me and said, "Son, you are to start at Willesden Garage in the morning, and don't forget 7.30 a.m." My consternation at this instruction can well be imagined when it is realised that I did not know the geographical location of Willesden Garage. However, after reading paragraph two on the copy of my indenture it was obvious that I had no alternative but to obey this instruction, which read:- "To learn in their depots situate at Green Street, Forest Gate, or elsewhere as the Company shall direct in the art, trade, or business of Fitter and Turner so far as the same relates to the said undertaking of the Company".

I was sorry to leave Leyton Garage. It was reasonably near home for me, and bearing in mind that I had been transferred from Forest Gate Garage already, I had cherished the hope that my apprentice-ship would be served and indeed completed at this establishment. The staff were very friendly and helpful, and of a somewhat younger average age than those at Forest Gate. The age-long leg pulls were practised with great frivolity, and sending new would-be engineers to to the stores for a "long wait" or a "left-handed clout" was part of the initiation to the staff. I had considered it a great advantage to me that there was only one other apprentice on the staff, and conse-quently we enjoyed greater freedom in movement from learning one task and then moving on to an entirely new job.

I must confess that we apprentices were not hesitant to remind the foreman (very politely) of our status, and to further our claim to be moved to another task. There was no fixed schedule of movements for apprentices at the depots, and the tasks allocated to them largely depended on the foreman, prompted mainly by the keenness of the apprentice to be moved and consequently to gain more know-ledge and experience. In general, if the foreman could see that lads were really keen on gaining experience in their craft he would usually assist all that he possibly could. Some of the fitters regarded the apprentices as only a necessary nuisance, and recognised them only when they wanted a dirty or disagreeable task undertaken, but most of them were very good and indeed would go out of their way to assist or help.

Quarterly, and subsequently annually, reports of the progress of indentured apprentices, signed by the Chief Engineer of the Company, were sent by post to the parents or guardians regularly.

23-43-L.G.O. Indentured Apprentice's Report.

LONDON GENERAL OMNIBUS COMPANY LIMITED.

ENGINEERING DEPARTMENT.

INDENTURED APPRENTICES.

Name of Apprentice Scanlon H.D.   Working at LEYTON GARAGE.

ANNUAL
QUARTERLY REPORT.
QUARTER ENDING 31.12. 19 20.
YEAR

Full Name	HENRY DESMOND SCANLON.
Home Address	134, Halley Rd, Forest Gate.
Age	16
Date commenced with Co.	8.8.18.
Commencing rate of pay	2½d
Present rate of pay	2⅞d and 2¾d W.W.

Diligence	Good
Adaptability to work	"
Initiative	"
Accuracy	Very fair
Quickness	Good
Courtesy, Cheerfulness and Co-operation	"
Class of work engaged on during year	Rear axles.
General attitude towards the work	Very good
Timekeeping	" "
No. of hours absent during year with leave	Nil
No. of hours absent during year without leave	"
No. of hours overtime worked during year	"
Absence through sickness, nature of complaint	" No. of days absent Nil.

ATTENDANCE AT TECHNICAL CLASSES.

Institute where attending	Poplar Engineering School L.C.C.
Subjects taken up	Machine drawing, workshop practice.
Days of week on which classes are held	Mondays, Tuesdays, Thursdays.

GENERAL REMARKS.

Indentured 30.9.20. Making good progress.

Date 1.1. 19 21.   (Signed) [signature] Shaw
District Depot Engineer.

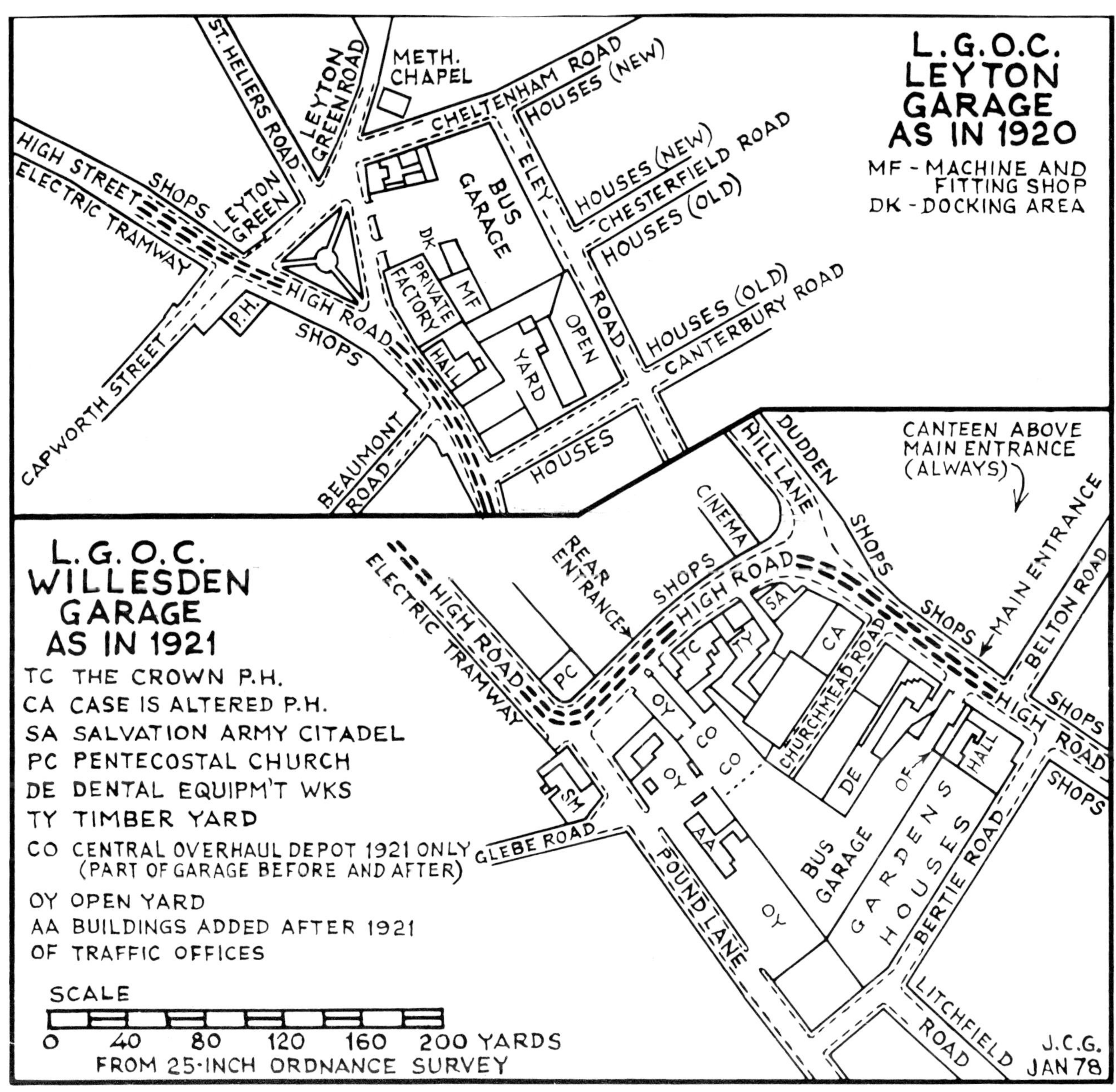

L.G.O.C.
LEYTON
GARAGE
AS IN 1920
MF - MACHINE AND FITTING SHOP
DK - DOCKING AREA

METH. CHAPEL
ST. HELIERS ROAD
LEYTON GREEN ROAD
CHELTENHAM ROAD
HOUSES (NEW)
HOUSES (NEW)
CHESTERFIELD ROAD
HOUSES (OLD)
HIGH STREET
ELECTRIC TRAMWAY
SHOPS
LEYTON GREEN
BUS GARAGE
ELEY ROAD
HOUSES (OLD)
CANTERBURY ROAD
P.H.
HIGH ROAD
DK
MF
PRIVATE FACTORY HALL
OPEN YARD
CAPWORTH STREET
SHOPS
BEAUMONT ROAD
HOUSES

CANTEEN ABOVE MAIN ENTRANCE (ALWAYS)

L.G.O.C.
WILLESDEN
GARAGE
AS IN 1921
TC   THE CROWN P.H.
CA   CASE IS ALTERED P.H.
SA   SALVATION ARMY CITADEL
PC   PENTECOSTAL CHURCH
DE   DENTAL EQUIPM'T WKS
TY   TIMBER YARD
CO   CENTRAL OVERHAUL DEPOT 1921 ONLY
      (PART OF GARAGE BEFORE AND AFTER)
OY   OPEN YARD
AA   BUILDINGS ADDED AFTER 1921
OF   TRAFFIC OFFICES

SCALE
0   40   80   120   160   200 YARDS
FROM 25-INCH ORDNANCE SURVEY

DUDDEN HILL LANE
CINEMA
SHOPS
HIGH ROAD
SHOPS
MAIN ENTRANCE
BELTON ROAD
HIGH ROAD
SHOPS
REAR ENTRANCE
HIGH ROAD
ELECTRIC TRAMWAY
PC
TC
SA
TY
CA
CHURCHMEAD ROAD
OY
CO
CO
DE
OF
HALL
GLEBE ROAD
SM
AA
POUND LANE
OY
BUS GARAGE
GARDENS
HOUSES
BERTIE ROAD
LITCHFIELD ROAD
J.C.G. JAN 78

My parents noted my reports with keen interest, and I was called by them to explain any further details about my progress in my apprenticeship which perchance was not mentioned in my reports.

Central overhaul depots, 1921

In preparation for the opening of the new overhaul factory at Chiswick, the London General Omnibus Company introduced four new Central Depots early in 1921 for the overhaul of the mechanical units and the chassis. These were situated at Hounslow, Willesden, Cricklewood, and Dalston, and were in fact in garages already operated by the Company, being really just the existing localised workshop areas now greatly extended. The Company at this time operated a fleet of about 3,000 vehicles, the overhaul of which was formerly undertaken in each of about 34 garages situated all over London, and so the idea of Chiswick Works was to centralise and standardise all the overhaul work in one establishment.

Staff from the various other garages engaged on overhaul work were gradually transferred to one of those four big depots. But consideration was not given as to where the employees resided; for example all the staff from Seven Kings garage were sent to Hounslow, likewise all from Leyton were sent to Willesden, and my former workmates from Forest Gate garage all went to Dalston.

It must be appreciated that in those days the staff were not informed officially of any plans or change of policy of the Company. On rare occasions any reference to changes reported in the Press would initiate all sorts of rumours. There was in circulation at the garages a single folded sheet under the title of "News of T.O.T. (Trams, Omnibuses, Tubes)", which was published by the Underground group of companies each month, and a very limited number were given away free. This news sheet was first published in 1914, and continued in circulation until the late twenties. Unfortunately most of the projects and schemes were reported, and photographs published, after the completion of the work, rather than before, and hence we poor souls at Leyton and elsewhere had no knowledge beforehand as to what was going to happen in the near future.

And so I had to leave Leyton Garage with great regret, conscious of the fact that working reasonably near home for me was finished, and I was to be faced with all the problems associated with travelling a long distance across the Metropolis in the very early morning. I was also very dubious as to whether the centralising of the overhaul work would in any way be beneficial to me in the process of learning my trade.

L.G.C.O. Willesden garage, 1921

Willesden Garage was built by the L.G.O.C. in 1912 specifically to house motor buses, which at that time were getting well established as the modern means of transport. It was on the south side of Willesden High Road, just east of Dudden Hill Lane. The front elevation of the premises was on the same building line as the adjacent property, with a rather narrow pavement, and being situated on a busy road this made entry and exit for vehicles very difficult. The building followed the same general layout of these two garages previously described in my story, with one important exception, which was that over the main entrance it boasted a canteen, which was controlled by the company. A small cup of tea could be purchased for a penny, or a large cup for three half-pence. Light refreshments and dinners could also be obtained at very reasonable prices. This facility was certainly much appreciated by newcomers to the garage, and was well used.

It was to this garage, which was being established as one of the four Central Depots, that I was transferred early in April 1921. It was also to here that "engineers", skilled and semi-skilled, were being sent at this time to undertake the overhaul work in an entirely different manner, which would eventually emerge ready for the complete progressive mass-production system which the opening of Chiswick Works was to herald. Willesden garage, although we could not really feel its buses were "our" buses like we did at Forest Gate and Leyton, had the code letter "AC" and worked at this time on routes 6, 8, 18, 36, and 46, all five of which remained substantially the same for the next 50 years or more.

I shall never forget my first day at Willesden Garage. It was my 17th birthday. I left home at 5.30 in the morning, and after paying my fare on tram and train (at workman's rate) I arrived just on time at 7.30 a.m. The first bus on route 25 did not start out from Forest Gate garage until six o'clock, so I could not use my "sticky", and I had to travel by L.C.C. or West or East Ham tramcar to Aldgate and then by Metropolitan Railway train to Dollis Hill via Baker Street. Bus passes were not then available on trains, not even when travelling to and from work, and this concession was not granted until 1941. No travelling allowance in either cash or time was given to any members of the staff who were transferred from one establishment to another.

I was full of apprehension as to what the garage would be like, never having been there before, and I wondered whether being an indentured apprentice to a large public transport undertaking was an advantage after all. On my arrival I was assigned to the engine final-assembly section, preparing the engines for testing on the test beds.

(Above) The main entrance to Willesden garage, photographed in 1926. The staff canteen was above the bus doorways.

(Above right) Canteens were becoming established as a feature of LGOC bus garages. This view taken at Seven Kings garage appeared in 'Bus & Coach' in 1929.

(Right) Women cleaners busy on B1260 in Willesden garage at about the time the author worked there. The vehicle carried destination boards for route 18, one of the main routes served by this garage.

The four-cylinder petrol engine of the B-type bus on which the author worked at Willesden garage. The cylinders, cast in pairs, were mounted on the aluminium crankcase. Prominent in the left foreground is the magneto, driven from the timing case and held in place by a metal strap.

This procedure was a foretaste of what mass-production assembly would be like at Chiswick, and I was not very favourably impressed. I was thankful that I was an apprentice, and that my stay on that account would be very brief. It was very difficult for skilled engine fitters who for years had built engines entirely on their own to adjust themselves to these quite short repetition tasks. It should be stated that at these four depots the units in the various stages of assembly were transported by hand. Various machines had been requisitioned from other garages and installed in the enlarged machine shop to cater for the additional work. This was a very difficult time for the staff, who for the most part had previously worked at a local garage, to be, at a moment's notice, transferred to an establishment an hour or two's journey away from home.

After a week or two on engine assembly I was transferred to the engine cylinder bench. These B-type cylinders were in blocks of two, with side valves (inlet near-side and exhaust off-side), and no detachable head. Cylinders were completely stripped and examined, but reboring was not undertaken at the depot. New valve guides were fitted, and valve seatings were cut by hand-operated cone and flat cutters. There were no valve seat inserts, as these were not introduced until 1930. All studs, plugs, and valve parts were checked, and afterwards the cylinders were tested for water leaks. This was accomplished in a very primitive, but nevertheless effective, method in a paved yard where the toilets were situated. This yard was not roofed, so consequently the testing of cylinders was temporarily deferred during rainy weather. The outlets of the cylinders were plugged, and water was hand-pumped to a predetermined pressure, which was maintained while the whole unit was examined for leaks. Gold size was used as a seal on all the water joints and water hose connections. After the cylinder blocks had passed this water test they were taken back to the bench for the valves to be fitted and ground, prior to the whole unit being assembled ready for fitting to the crankcase.

After working for some weeks on engine cylinder blocks I was transferred to the shop on engine crankcases, under the direction of a fitter. We had to complete one engine case, including its crankshaft, in one day. The engine main and big-end bearings were fitted and scraped by hand, blue marking on the crankshaft being used to determine the contact of surfaces. All the bolts on main and big-end bearings were drilled, and slotted nuts were fitted and subsequently split-pinned. The engine cases were mounted on timber trolleys, and when one fitter had completed his day's work the trolley would be moved to the neighbouring fitter, who proceeded with the next series of operations, these in turn amounting to a normal day's work for him. Apprentices were given instruction by the fitters, and after a period would be moved from one series of tasks to another.

When finishing time was approaching, both for dinner break and at night, the foreman would proceed to the entrance of the workshop area, with watch and whistle in hand. At exactly twelve o'clock for dinner, and five o'clock for finishing time, he would blow the whistle. This would be the signal for a general stampede to the garage entrance to queue up to "clock off". The eagerness of the employees could in some measure be understood, especially at night, when many of them were faced with a long journey home.

To those of us who had been transferred from a garage reasonably near to our homes this regimentation and exceptional rush was certainly foreign. The "family spirit" which had prevailed at the garages whence we had been transferred was completely lost, and we were now in an establishment which was neither a garage nor a factory. The whole atmosphere of the place was typical of a temporary organisation set up during an interim period awaiting the completion of the building of Chiswick Works. It was an absurd position! We employees knew little, we were told nothing. In those days advance information was not given to the employees of any intended changes, however much it affected them personally, and not even to indentured apprentices who were bound by contracts. It was ridiculous that a notice could not have been posted on the time recording clocks, giving employees some enlightenment on what changes were likely to take place.

For the remainder of the time which I spent at Willesden depot I was moved from various tasks around the fitting shop until I eventually ended up operating a lathe. This was very interesting work, for I undertook any job which came along — from chasing threads on brass engine-valve caps by hand-chaser, to screw-cutting on the lathe the thread on special bolts or screws which from time to time I was called upon to make. This meant changing the gear wheels in the train of gears at the end of the lathe bed. A clutch was not fitted, hence when engaged on screw-cutting or tasks where a very definite stop was necessary, for the last one or two chuck revolutions required the power would have to be shut off, and the spindle revolved by pulling the belt by hand. There was no self-act shaft fitted to this machine. To traverse the saddle along the bed by power it was necessary to utilise the lead screw. Engine valves were turned, and the finish which was demanded was almost equal to a ground finish.

We had by now received word through the grape vine which has operated in workshops since time immemorial that the transfer of at least some of the staff to Chiswick was to take place in the very near future. Of course there was speculation as to whether the "blue-eyed boys" would be the first to go. Most of us felt that we would welcome the transfer to be at least in some measure settled.

My time at Willesden was terminated even sooner than I had anticipated. As on the last occasion that I had been transferred, very much the same pattern was followed. One morning in September 1921 at about ten o'clock, the foreman approached me and said "Son, pack up all your tools in your box, mark it with your name, and put it on that lorry" (pointing to same). He then continued, "Get your coat on and take a seat on the bus behind the lorry, which is to take you to Chiswick Works". We were subsequently to learn that the staff were to be gradually sent to Chiswick in reasonably small groups. With me on the bus en route were about another 20 members of the staff. Incidentally the pattern and dimensions of the tool boxes used at this period are identical to those still in use at the garages today.

So we were taken in this bus to Chiswick Works, with the lorry containing our tools following, and thus started the exodus which ended my career at Willesden Central Depot. This period had not been a particularly happy time for me, working in an atmosphere of uncertainty, using makeshift tools, adopting temporary methods and procedures in the execution of one's allotted tasks, and constantly being told by the different foremen that "things will be different at

Bus-washing, 1922 style. This gantry was used to spray water on buses at Willesden garage. The B-type bus operated on service 6, which is still worked by Willesden. The notice on the wall in the background warned drivers that they were liable to instant dismissal for travelling at excessive speed.

Chiswick". I have often reflected as to in what channel ran the minds of the individuals who controlled our destiny as to where we worked and when we were to be transferred. I had no idea that I was to be moved on this particular day, and hence had purchased a return workman's ticket on the railway ready for my homeward journey. Being now, to my dismay, transferred to Chiswick, where I had to remain till finishing time, I had a useless (to me) return-half railway ticket on a different route. In actual fact, to avoid further expense, I went home all the way by service bus with my "sticky", which took two hours, and as this was an entirely new route to me I also had to make enquiries to find out how and where to do it.

3. Early Days at Chiswick

Inauguration of Chiswick Works

The building of the new Chiswick Works had been started in September 1920, on a large plot of ground that had hitherto been used for growing rhubarb, cabbages, potatoes, etc. The site was on the north side of Chiswick High Road, opposite Gunnersbury Station, but of a rather irregular shape, roughly rectangular, bounded on three sides by three different railways and on the west by Silver Crescent. Only one corner of the site had a frontage on to the High Road itself, but the main entrance was situated here, hence the entire establishment carried the official postal address of 566 High Road, Chiswick, W.4. Previously, while I was still working at Leyton and Willesden garage, the workshop staffs there always referred to this new project about which we knew so little as "Gunnersbury Works", presumably because the railway station of that name was so close. Indeed this could well have been a more correct geographical description, because the site was beyond the outer edge of what was then the continuously built-up area of residential Chiswick.

By March of the following year sufficient of the building had been erected to enable a certain amount of work to be turned out in the coach-building half of the factory. The previous main coach-building factory in North Road (off Caledonian Road), also the two subsidiary ones in Olaf Street (off Latimer Road) and Seagrave Road (off Lillie Road, Fulham), were all closed down later in 1921. The engineering side of Chiswick did not begin operations until the month before I arrived, August 1921, but from then onwards contingents of staff from the four Central Depots were being gradually transferred to Chiswick Works to inaugurate the various mechanical sections. This process continued until May 1922, by which time all the staff had been transferred, and the Central Depots as such were closed and the premises reverted to normal garage routine. It is interesting to note that today, more than 50 years later, although London Transport has now closed quite a lot of other earlier L.G.O.C. garages, these four are all still flourishing. Hounslow has been rebuilt, but Cricklewood, Dalston, and Willesden are all substantially unchanged apart from numerous detail improvements.

The official opening ceremony of the new Chiswick Works was held on 31st January, 1922. The general layout of the factory was divided into three parts; thus the Eastern end was the Engineers, the Western end was the Coach Factory, and the central part was mainly the Stores but included also the Experimental Shop, the First Aid room, and the fire station. All these sections were under one

(Above) The main entrance to Chiswick Works in the early days. This photograph was taken around 1923-4, three or four years after it was opened. (Below) It has not greatly altered over the years — the view below was taken in 1975.

extended covered roof, running east to west in seven bays, while encircling the whole factory a test road with a rough surface was built as part of the general scheme. This involved the making of over 800 ft. of graded roadway, with a double test hill, known as "The Dip", on its south side. From the excavation required to construct The Dip all the sand and ballast needed for the building of the works was obtained. During World War Two The Dip was kept permanently flooded, to provide a very useful emergency water supply for fire-fighting.

There was a well-equipped First Aid Station, with a fully certificated nurse in attendance, and a motor ambulance always ready to convey any possible serious casualty to hospital. Isolated from the main factory, near one end of The Dip, there was a separate building of large capacity equipped as a Canteen and capable of seating over 1,000 people. When not required for the ordinary daily purpose this could be used for every class of entertainment, and numerous

"The Dip" at Chiswick Works. This view of the test hill was reproduced in the first issue of 'Bus & Coach', January 1929 and shows NS 1623, built in the mid-twenties as an open-top bus but by then converted to covered top.

A lunchtime view of the main exit roadway in Chiswick Works in October 1933, showing employees who lived nearby walking towards the main exit, homeward bound for lunch. The prominent building with the clock tower is the canteen.

concerts, dances, children's parties, amateur theatricals, boxing matches, etc. have been held here in the evenings. Another separate building, alongside the canteen but on the opposite side of the main roadway which led up from the main gate, housed the administrative offices. All the original buildings in the whole establishment were of single-storey type, built only at ground-floor level, but an additional floor was added to the office block in 1931.

Five separate and smaller buildings were added within the first ten years along the northern edge of the main factory, but separated from it by the Test Roadway. These housed the Welding Shop, the Millwrights, the Timber Stores, the Boiler House, and the stores for obsolete material. Down near the main gate, and not really part of the bus overhaul factory at all, the Ticket Section established their headquarters, for the repair of ticket punches and the despatch of bus tickets to all the garages. Another newer addition, at the eastern end of the site alongside the District Railway, housed a greatly enlarged Experimental Shop and the Fire Station, which now moved out from the original building, and also the uniform Clothing Stores. Here also was the original skid patch, until 1938.

A further new building, somewhat higher than any of the others, was erected in 1938 alongside the eastern end of the main Engineer's section. This comprised three galleries as well as the main ground floor, and most of it now became the General Stores for new material, but the southernmost end housed the engineering view room of the Inspection Section, and the northernmost end was devoted to Wheels and Tyres. The original General Stores was now closed, and its space was taken over as enlargements of both the Engineers and the Coach Factory. Another addition, also in 1938, was a separate new building on the opposite side of The Dip, and to reach it a road bridge was built across the deepest part of The Dip. This building contained an enlarged Engine Assembly section, and an enlarged "Testing and Rectification" section for checking the finished overhauled and reassembled chassis. A very well-equipped Training School for drivers and conductors was built opposite the Ticket Section just before the war, and enlarged to have a second storey soon after the war. Finally, an additional South Wing was added to the main office block in 1956, again of two storeys.

In the early days of Chiswick it was necessary, under a Metropolitan Police Force Regulation, to carry out vehicle overhauls annually, and within a few years the output had risen to 17 overhauled vehicles per day, or one every half-hour. By 1939 the bus and coach fleet had increased to 6,000, and these were vehicles of a much more complicated design and construction than the earlier B and K types. However, amendments to the law had now made it no longer compulsory to carry out vehicle overhauls annually, and on average the vehicles at this time ran in service for about 18 months between overhauls.

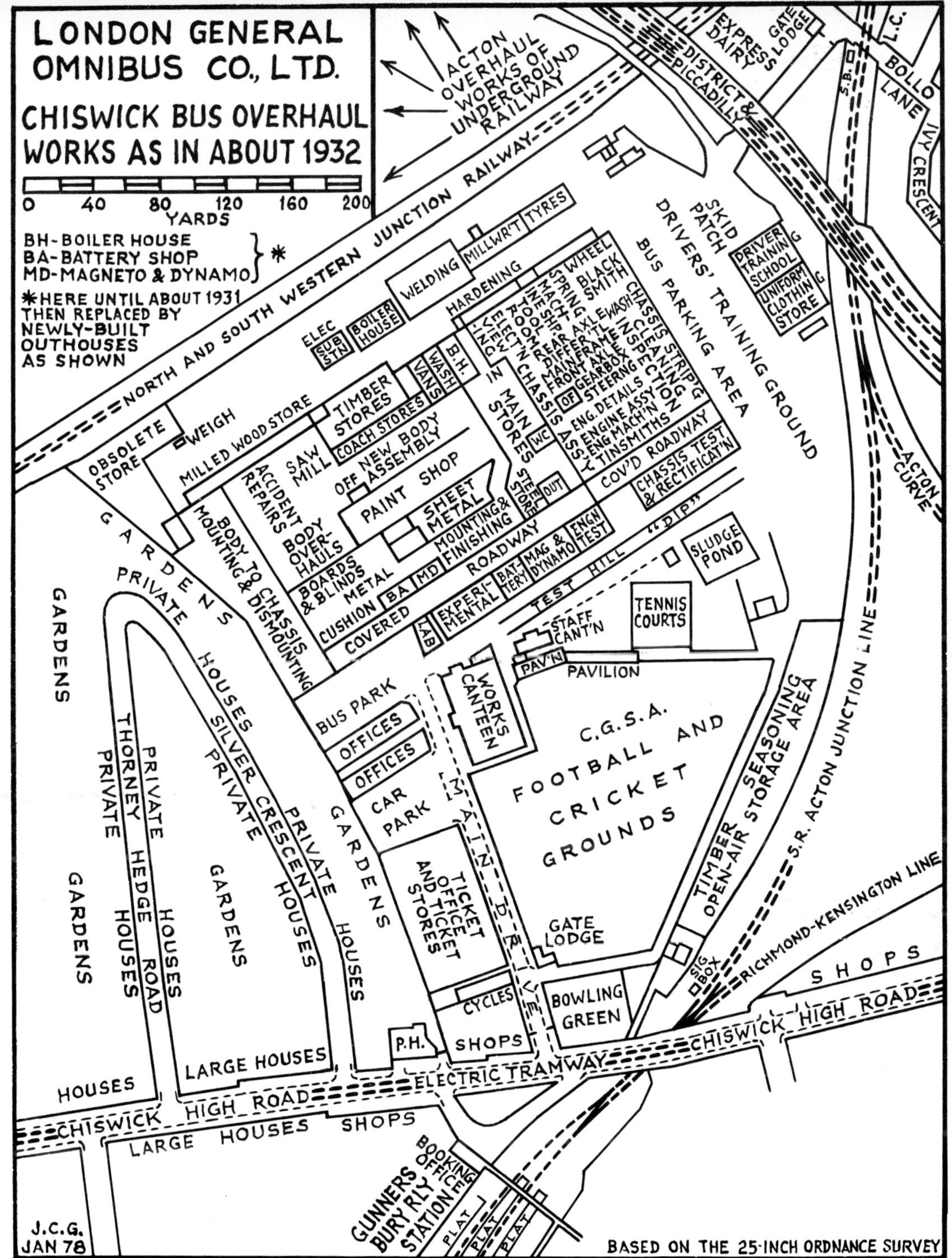

This plan shows how Chiswick Works was laid out during the earlier years the author worked there. Later additions are shown on page 82

My arrival at Chiswick Works

On arrival at Chiswick Works from Willesden Garage in September 1921 we were told to line up outside a certain corner office, which we were later to learn was called the Labour Bureau. Here, we had to answer numerous questions which were recorded by a clerk on an official-looking form, then we were allocated a clock number, and told to sign the form on the dotted line. I always remember I was severely reprimanded for asking to be allowed to read what I was being requested to sign. The whole procedure I am sure was similar to the reception given to prisoners on being received into prison. We were then told to follow a "gentleman", who we afterwards found out was to be our foreman and who escorted us to the bench where we were to work, but an introduction to him was not given. When we wanted to know anything we had to seek our own salvation, and find the answer the best way we could.

However, here we were in a new factory, only three-parts finished, the site of which only 12 months previously had been a market garden and rhubarb field. The section to which I was allocated was engaged on details such as change-speed and compensating gear, and was always known as the Detail Section. The work consisted mainly of rebushing and reamering the setting arms and brackets. It was quite interesting work at first, but very soon got monotonous. The finished details were not inspected, but it was very important that they were marked with chalk by the progress clerk to indicate that they had been counted and recorded as output for the section. They were then transported on hand trucks to the chassis assembly line.

The most striking feature in the new factory, on the engineers' side, was the raised rectangular office block, the floor level of this block being about 6 ft. above the shop floor. It was used to accommodate the Section Engineers and their respective personal clerks. It was glazed in clear glass on all four sides, and was known to all of us as the "Glass House". From this building it was possible to view almost all of the production sections. The general policy prevailing in those days was for it to be possible for the supervisors to see all their staff working at their individual locations. The general rule was that cupboards for tools or equipment were not to be much above bench level, certainly not up to shoulder height, and this rule even applied to stocks of material or units which had been stacked awaiting repair.

Ridiculous as it may now seem, in those days if an employee was caught sitting down during working hours, even though he might have been gainfully employed and correctly pursuing his normal duties, he would be severely reprimanded, and if caught again in all probability discharged. In the whole establishment, with the exception of just a very few welders, and the sewing machinists in the trimming shop, who anyway could not operate their machines standing up, you could not find one single employee sitting down to do their work. Even the progress clerks had very high desks, and stood up to do their writing. It seems utterly stupid to relate now, but even if you thought you could make a better job by working sitting down, or could get better access, or see it more clearly, you still just simply had to stand! That was the RULE.

The signal for starting, dinner, and finishing times was given by the sounding of the works hooter. This was extremely loud, and could be easily heard from Chiswick Park railway station. It was used continuously until 1939, when it was converted and used as an air-raid siren, and was replaced by a series of electric bells. But these in turn have by now been replaced by "pips", similar to those broadcast on the radio.

Entering and leaving the establishment in 1921 was under the strict supervision of the staff of Wardens at the front and rear gates. On entering by the main front gate it was necessary for employees to show their bus travel pass, but if you were not employed by the company you had to state the reason for your visit, and were escorted to the person or persons you desired to contact. The rear gate was opened for staff to enter or leave only just before starting or finishing time, and to use this gate it was necessary to obtain a special pass. This pass, which was numbered, was issued to the holder only, and he was required to sign it. It was headed "Pass Out to South Acton Station via Bollo Lane Gate", and read:- "This Pass must be used by the holder only, and he must produce same upon demand by a Company's official". It was signed on the reverse side by the engineer-in-charge.

Personal packages and parcels, etc, were not under any circumstances allowed to be taken into the works, but had to be lodged in the canteen on special racks which were provided. An employee wishing to take articles out of the works which were his own property had to obtain a pass-out advice from his supervisor, which had to be handed to the warden at the works gate when leaving. The Company reserved the right for the wardens to examine any such personal packages and parcels, etc., when being taken off the premises.

Free travel passes on the London United Tramways, Limited, two of whose main routes ran along Chiswick High Road past the front gate, were issued to certain members of the works staff on application to the Labour Bureau. These were available from the stop at the works main entrance, as far as the tram termini at Shepherds Bush or Hammersmith, for use in either direction for journeys to and from work only. The all-routes free passes and duty

Chiswick High Road outside the LGOC works, looking west in about 1930, and showing London United tram No. 350 on one of the routes used by employees. This particular car was the only tram ever built in the Chiswick Works, in 1926, when it was supplied to Metropolitan Electric Tramways as their No. 139. Both MET and LUT were associated with LGOC as fellow-members of the Underground group but subsequent tram, and trolleybus, manufacture was concentrated in the works of the Union Construction Co. No. 350 shows clear evidence of its bus works origins in the appearance of the upper-deck, showing a marked resemblance to the NS bus following.

after only a couple of years were cut down to leave a gap at the bottom of about 18″ from ground level, with the top of the door only about 4′-6″ above the ground, and remained so for many years. I strongly suspect that this action was ordered by some petty official of the Company who resented the "ordinary" bus workers having any privacy while on company premises, though perhaps it also related to the "no smoking" rule mentioned later.

Employees' coats were put on racks hoisted up towards the roof during working hours, as was common practice in factories at the time. This photograph was taken in the Boards and Blinds Shop in January 1936. Destination boards were still in fairly widespread use, though blinds were taking over by then.

bus passes, for staff travel on L.G.O.C. buses, have already been outlined in an earlier chapter, and the old arrangements for these at the garages now continued unchanged at Chiswick. But these passes were not available on buses between the Chiswick Empire Theatre (now demolished and replaced by Empire House) and the corner of Gunnersbury Avenue (where the big roundabout and flyover now is), for travel in either direction, between the hours of 4.30 p.m. and 5.30 p.m. on Monday to Friday, or 11.30 a.m. to 12.30 p.m. on Saturday. Hence employees tended to walk rather than ride, for the first half-mile of their homeward journey. At this time it was also a rule of the company that not more than three pass holders were allowed to travel on any one service vehicle at the same time.

The feature of the works that probably impressed and pleased all of we newcomers the most was that accommodation for washing was provided, including soft soap and hot and cold water. This was something which we had not enjoyed at the depots. Barrier cream, however, like that issued to all works staff today, was not then issued. The toilets too, although situated in an open yard, were very much improved, and incorporated full-length doors, which, alas,

Coat hooks also were provided in the new factory, these usually being sited adjacent to the time-recording clocks. They were on racks accommodating eighteen hooks on each of two sides of a rectangular framework. These racks were supported by wire ropes, running on pulleys up to the main roof structure, and during working time they were hoisted up to the roof by hand winch, where they remained suspended until lunch time or finishing time.

A weekly issue of clean rag, mostly old and torn garments, was made to all fitters and machinists for general use in any cleaning job to be undertaken in the execution of their allotted tasks. Often various articles of women's underclothing would be included in this issue, and, as can be well imagined, would cause many a laugh and joke. This old rag when used would be collected, and clean ones issued in exchange, and the dirty ones returned to what was known as the "rag shop", which in fact was a laundry. The oil and grease extracted from the rags would be sold, and the rags used over again.

The winter of 1921-22 was very severe indeed, and unfortunately for those of us who had been transferred to the works the heating installation had not yet started to function. It was like working in a colossal partly finished and half-empty barn, and as can well be imagined it was extremely draughty. The management realised the conditions under which the staff were working, and gave instructions that braziers were to be introduced temporarily. These were old scrap rectangular water tanks, pierced with holes on all four sides and mounted on bricks about 1 ft. high. They created volumes of smoke when they were first lit, but, once well alight, served at least in some measure to raise the temperature inside the premises. The concrete floor had been laid only recently, and was so cold and damp that we had to stand on old scrap plywood panels taken from vehicles in the process of being overhauled. But as the progress of building advanced, so we were provided with slatted timber duckboards for all benches and machines, and these gave about an inch of clear air space between our feet and the concrete, helping to keep us a lot warmer.

In this post-first-war period the management of many industrial undertakings had begun to realise that employees could, and usually would, work harder if some system of heating was installed in the factory, thereby increasing their production and efficiency. In this regard the L.G.O.C. were no exception, and arrangements were already in hand for the installation of suitable heating and ventilating apparatus for Chiswick Works. This was designed to maintain a temperature of about 60°F throughout the building all the year round, and in hot weather the ventilation provided for the flow of fresh air into every part of the factory. To those of us who had been transferred from various depots, the prospect of working in a heated workshop was something new and indeed very welcome.

Machines, with their countershafts and power drives, were still being sited and permanently installed. At this period all this class of work was carried out by the Millwright Section, which in these modern times is called the Plant Department. A compressor plant had been installed in approximately the centre of the factory, and air lines were made available to the various sections to power hoists, or to air drills, which were known by all as "windy drills". This pneumatic equipment was something new to all of us, who had previously used only rope tackle or chain hoists for lifting. Drilling on the bench or on vehicles had previously been accomplished by ratchet drills, or wheel braces.

The new works was equipped with its own tool room, which dealt with the manufacture and maintenance of all special-purpose jigs, tools, and fixtures, for use at Chiswick and at the various garages. The tool stores was built adjacent to the tool room. Tools were loaned from here when a brass disc (of 1½″ dia. and 17 s.w.g. thickness, and embossed "L.G.O.Co. Ltd. Works", and stamped with the employee's clock number) was presented by him, with a verbal description of the tool required. This disc was retained by the storekeeper until the tool was returned. This system worked very satistorily, and while somewhat crude in character, it was nevertheless efficient. It would seem to the author that there is little reason to merit the present-day workshop practice of presenting a written tool request form, embracing a full description of the tool required to loan, signed and dated by the employee and countersigned by the foreman. This is a costly and surely unwarranted procedure, which only increases the volume of paper-work associated not only with bus work but also with most other modern businesses.

One of the greatest difficulties confronting the management in 1921, when the chassis and all its attendant units were being overhauled at various garages, as previously described in my chapter on Forest Gate Garage, was to obtain perfect cleanliness in parts that had been grimed and greased after many months of service. To overcome this problem, a special caustic soda washing machine was designed and built at Chiswick Works, and it proved to be a most successful project.

Before the layout of the works had been completed, two of these machines were installed and in operational use. One of them was used mainly for complete chassis frames, and could in later years easily cope with the larger LT and STL chassis although installed in the days of the smaller B and K. The washing machine at Chiswick today, in 1979, follows the same broad outline of these machines as first conceived.

Extensive use was always made of roller conveyors for moving the parts to and from the washing machine, or otherwise material was transported on hand trucks or sack barrows. Many years later Lister Auto-Trucks powered by air-cooled petrol engines were introduced, and these ultimately were superseded by electrically driven trucks very similar to those in use today.

Staff conditions at Chiswick

All employees engaged at Chiswick, whether transferred from other depots or new entrants, were issued with a Rules and Regulations Book, for which they had to sign on receipt. This book contained 41 Rules set out under various headings. While freely admitting that the vast majority of them were merely formal instructions covering the conditions of employment, two of them covered

The caustic soda frame washing machine, photographed for an article in Bus & Coach of December 1930. It was also used for cleaning large aluminium parts. Old paint was burned off before the frame entered the machine.

the two worst crimes it was possible to commit. One of these two crimes was to "fiddle" the clock, or in other words to clock in or clock off somebody else in his absence, or in any way to procure a false reading on any clock card. The penalty for this, if caught, was instant dismissal.

The other major crime was to be caught smoking on the premises. Large prominent white notices with red lettering were displayed in very many positions around the factory, warning employees not to smoke on the premises, and reminding them that if caught they were liable to instant dismissal. The rule was very rigidly enforced, and should any employee be caught smoking he just had to say good-bye and be off. And yet men did venture to smoke surreptitiously in the old days, ridiculous though it may seem, usually in hide-outs or the toilets. The management were well aware of this, and on occasions a raid would take place on the toilets, but the grape vine of the workshop would always be one jump ahead, and although the toilets would be found with their air thick with tobacco smoke no culprits would ever be found.

In the light of present-day conditions this no smoking rule would seem ludicrous, for nowadays smoking is allowed pretty well all over the works with only very few exceptions. Indeed batteries of cigarette slot machines are now installed in various locations in the workshops and the main office block. But the important difference, of course, between the old days and today is that nowadays we have oil engines and metal-framed bodywork instead of petrol engines and bodies built almost entirely of timber. In the 1905-20 period, when these rules were formulated in the light of experience, there had been many instances of buses catching fire inside garages while being worked on.

Here it would seem appropriate to mention the main staff code of the workshop. To "shop" one's fellow employee, i.e. to blab or tell tales, was an unforgivable sin, and simply was not done. If the management discovered that some irregularity was being pursued, and certain employees came under suspicion and were being questioned regarding same, in all such cases the necessary evidence required from other men in the shop to convict the person or persons simply could not be obtained, because everybody stuck together to protect their mates.

The canteen, to those of us who were transferred to Chiswick in the very early days, seemed a very large building, and certainly a great contrast to the small canteens at the various garages and depots whence we had come. It was equipped with trestle tables covered with black-and-white chequered lino, and with timber chairs which we called kitchen chairs. It was a palace to us, and quite warm in the winter. A good hot dinner could be purchased for a shilling or less. Meat dishes were sixpence, and various vegetables were one penny a portion, while sweets including custard were two-pence half-penny. A small cup of tea could be purchased for one penny and a large one for three half-pence. These dinners were very good in quantity and quality, and there was far more variety then than there is today. On the whole they represented very good value for money. Friday was always fish day, when fish-and-chips could be obtained for eight-pence, but of course other dishes were also available if you preferred them.

There was no licence for the sale of intoxicating drinks at first, although this was applied for and granted many years later. In the meantime beer and spirit drinkers had to resort to the John Bull public-house in the High Road just outside. During 1924 certain alterations were made in the canteen, which made it possible for a film show to be given every Wednesday lunchtime, starting at twenty minutes past twelve, and lasting for half an hour. These shows were extremely popular, regardless of the fact that the films were old ones. For the meals the employees from the workshops used the main canteen hall, while office staff and supervisors were accommodated in a section (it is now embodied in the kitchens) which was called the "glass house" because one whole side of this structure was entirely glass. But this must not be confused with the other "glass house" mentioned in the previous chapter, the raised office block in the factory.

It is interesting to note that the office staff were known as from "over the road" this being the location of the office block in relation to the workshops. Similarly the head office staff from 55 Broadway were known as from "up the road". Any trouble arising on the shop floor that could not be settled by the supervisors was known as going "over the road", in an endeavour to reach agreement, and likewise any serious trouble affecting the personnel of the whole works had to be settled at head office by the "heads up the road".

Operating at this time there were quite a number of sick and loan clubs run by individual employees in the workshops. The members would pay a shilling or two each week, and a share-out of club funds would be made at Christmas. These clubs served a very useful purpose, for at this period employees were not paid for holidays, and a share-out at Christmas helped in some measure to compensate for loss of wages caused by the enforced holiday. Unfortunately on two separate occasions when the time for the pay-out arrived the club treasurers concerned found they were considerably short of cash, having had the club funds "mixed up" with their own money, and so they subsequently absconded with what was left. This came to the notice of the management, who promptly and wisely made arrangements for moneys in future to be collected and paid in at the works cashiers, and a more stringent arrangement for the handling of such funds was introduced.

The idea of a Chiswick Works staff outing was debated in the very early days of the opening of the works, but nothing concrete transpired until June, 1923, when an outing was arranged to take place one Saturday with Blackpool as the venue. An entire train was chartered to transport the party to the seaside by railway, and liberal supplies of liquid refreshment were made available on the train to sustain the travellers. The works were closed for this Saturday, and to save the loss of wages caused by this the management by agree-ment allowed two hours of optional overtime to be worked on each of two evenings preceding the event. It must be realised that at this period a paid holiday was not yet in being. This outing, usually to Blackpool, and the accompanying closure of the works, was an annual affair for many years, and much enjoyed by all, but it was discontinued shortly before the start of the Second World War.

Stocktaking was undertaken in the workshops at Chiswick at Easter of 1922 and 1923. This function, previously undertaken at the various garages and depots which Chiswick Works had replaced, had not then affected the staff in any way. But unfortunately at Chiswick this stocktaking operation meant the works being closed for production for a whole week. This was a financial calamity for the hourly paid staff, for it meant no work for a week and therefore no wages. It took the staff many weeks to make up this enforced idleness, and one can easily imagine the distress that the stocktaking operation left in its wake, and so the entire personnel of the workshops were indeed thankful when after two years this practice was discontinued.

What replaced this former procedure nobody knew or cared; it was sufficient to know that the heartless and distressing closure of the factory was at an end. In those days Bank and statutory holidays were in effect a lock-out to the workshops staff, or an enforced holiday. How anyone could designate a period when men were shut out from work without pay as a holiday revealed the callous in-difference to staff relations at that time, which was apparent not only at Chiswick but over industry as a whole and throughout the country. To most of the personnel these holidays were a financial embarrassment, indeed a calamity, which took many weeks before a reasonable financial recovery was possible.

In 1922 the British Broadcasting Company was formed (it became a Corporation some years later), and the 2LO London radio pro-grammes first started broadcasting. It is quite surprising what a great benefit we soon gained from this. Irrespective of the knowledge and entertainment one big advantage was that we now had something reliably accurate by which to set our watches and clocks. Until then we had had to rely solely on public clocks for getting the correct time. We at home would go into the garden and listen for the clock at the local public library to strike the hour, and then check our watches and clocks from this. Other people elsewhere would similarly use the public clock on their local Town Hall or parish church, but of course many thousands of people had no public clock at all that was conveniently local to them.

New models introduced by LGOC during the author's apprenticeship days included the S-type, which was an enlarged version of the K-type, capable of seating up to 54 passengers. The first vehicle was built in 1920 and 849 double-deckers and 79 single-deckers were built, mostly in the period up to 1923. The four vehicles shown, S198, 244, 245 and 85, were operated on the world-famous 11 route from Hammersmith garage.

To know the right time was absolutely essential to us hourly-paid bus workers, both at the garage and also subsequently at Chiswick, because the control of timekeeping was always extremely strict. We were allowed just two minutes' grace, but if we arrived for work only one minute later this caused the loss of half an hour's pay. Not that the L.G.O.C. was any worse than elsewhere, for this was indeed the normal practice in most branches of industry. Not until the 1940's was the position slightly eased, in that from then onwards we had only a quarter of an hour's money deducted from our wage packet for being one minute late. And so you can see that the new B.B.C. time checks in 1922 were an enormous help to us. In those days absolutely everybody, including certainly all the garage and factory workers, wore a waistcoat and a watch and chain. Most working people wore an Ingersoll watch which could be bought for 1/11d, but those a little higher up the social scale wore a stronger Swiss-made gun-metal watch costing about 3/11d. The watch-chain was rather a status symbol, and those who could afford it, or who had generous relatives at Christmas time, wore a chain made of solid gold.

Prior also to the coming of radio, when major horse races were being run somebody in one of the offices would receive the result over the phone, and then it would subsequently "fly" like lightning around the workshops by word of mouth. Radio broadcasting of major horse races soon altered all this, for just prior to the race being run our crystal sets would be manned in the toilets or other hide-outs and the result was gained that much earlier. I'm not sure now how we managed to dodge the foreman! After the Second World War the management allowed the Derby to be officially relayed over the works loud-speaker system which had been previously installed during the war to relay the "Music While You Work" programmes. They evidently took the view (and I would heartily agree with them), that this action on their part saved a lot of running about by the staff and the consequent loss of production.

The Chiswick 'General' Sports Association

Early in 1923 a Sports Club for employees was formed, and was named under the title of the Chiswick "General" Sports Association, or C.G.S.A. for short. The subscription for members was fixed at two pence per week, payable through the company's pay rolls, and deducted each week from wages due. For this small sum the members were entitled to participate in all activities sponsored by the association. The great majority of the staff joined this club, which was indeed good value for money. Football, swimming, boxing, rowing, and athletics sections were formed, and very soon began to flourish, and, as time progressed, so various other sections were inaugurated. Various traders in business in Chiswick High Road quickly realised the trading potential of such a large sports club, and allowed a discount to members on purchases in their establishments on the production of their sports club membership card.

By this time the Sports Ground adjacent to the factory had been completed, and grass tennis courts laid out, which were made available to members for matches or practice. The first athletic sports meeting was held on the sports ground in July 1923. The programme, consisting of over 24 events, was a great success, and thereafter was an annual event until the whole sports ground was taken over in 1938 for the erection of various buildings and the construction of the driving-instruction skid patch. A keen interest was shared by all in the achievements of various members of the staff on the athletic field.

During 1926 I was privileged and indeed honoured to be chosen to run the half-mile leg in the Engine Section and Works relay teams. The Athletic Section of the club consisted of a grand crowd of fellows who were really keen and enthusiastic. We would train on one or two evenings a week during the spring and summer, on the sports ground after finishing work at five o'clock. There were excellent dressing-rooms and showers available for our use, which were situated under the grandstand.

I shall never forget running the half-mile leg in the one-mile relay team representing Chiswick Works at the Busmen's Sports held at Stamford Bridge. It was one Thursday afternoon in August 1926. I was up at five o'clock that morning, and journeyed from home to be at the works at 7.30 a.m. I stood at my machine, on crankshaft grinding, until 3.30 p.m., and then travelled with the rest of the team by bus to Stamford Bridge ground. We all had special permission from the Labour Superintendent to leave work early, but without pay, so as to be at the ground just prior to the time our event was scheduled to be run. We were all conscious that to waste time at the ground would have cost us even more lost time and lost cash. Surprising as it may seem, the half-mile I ran on this occasion was my best ever for this distance, in just over two minutes, a time which I was never able to repeat. And we won the race, thanks to the grand support of the other members of our team.

At the fifth Annual Sports day held at Chiswick Works in July 1927 an innovation was tried. A film unit was in attendance, taking the high-lights of all the afternoon events. It was well advertised that the film would be shown later, as an extra feature at the Chiswick Empire Theatre during the performances throughout the following week. The Empire, which was demolished in the early sixties, used to stand in Chiswick High Road opposite Turnham

(Above) The author leading the field at the Chiswick General sports day, 10th July 1926, held on the sports ground within the works. In the background can be seen, from left to right, the ticket office, the main office block and the staff canteen.

(Left) Contemporary recognition.

(Below) The author breaking the tape in the 880 yards handicap race at the 1927 Chiswick General sports day, on 9th July of that year.

Green, on the site where the Empire House office-block now is. I saw the film, and had the unique experience of seeing myself running, and winning the 880-yds Members' Handicap Race. Whether the film was a financial success or not I cannot say, but this was the only occasion that a sports meeting held on the works ground was filmed. In more recent years, however, the annual boxing tournament held in the works canteen has been televised by the B.B.C.

In the early years of the works there was a bowling green on the east side of the works entrance, just inside from the Chiswick High Road. This attracted very many of the staff, and numerous bowls competitions were held here. Indoor games also were well catered for, for these as well as whist drives and dances were held regularly in the canteen. So also was the annual Children's Christmas Party, and indeed this party is still held each year.

In the old days a far keener interest in amateur sport was apparent in large industrial concerns such as Chiswick Works than is the case at the present time. This interest has now been very largely replaced by professional football, coupled with a very real and lively financial interest in the football pools, which are quite a modern innovation.

Back in the twenties and thirties it was a most serious offence, punishable with instant dismissal under Rule 29, for the staff to take bets in the factory, even in their own time. Nowadays there are two betting shops within a stone's throw of the main entrance.

Safety first and health hazards

In those early days of the opening of Chiswick Works the general principles of Safety First, and the use of machine guards, while basically being observed were nevertheless extremely haphazard. However, to be quite fair, this was very general in all industrial undertakings at this time. Industrial eye protectors were simply not thought necessary at this period, and indeed were not available. Nowadays, the paramount importance of having proper eye protection for all classes of work has been fully recognised, and protection *must* be provided by law under the Factories Act of 1937. Dry grinding in the old days was a very common operation on many machines, but unfortunately dust extractors were not fitted.

One personal experience of this will for ever be in my memory. Soon after finishing my apprenticeship in 1925 I was assigned to a job of boring B-type cylinder blocks on a purpose-built machine. This was done by two rams fitted with revolving multi-cutter heads which had been made in the works Tool Room. After the machining operation we had to sweep the fine cast-iron dust away from the cylinder bores, and we could taste it in our mouths. Heaven knows how much penetrated into our lungs, if the amount that got grimed into our clothes was any indication. The final finish of the bores was accomplished with a portable pneumatic drill, fitted with a buff which had previously had carborundum powder glued onto it. It can easily be visualised that with this last operation the volume of dust in the air was greatly increased.

This job and its conditions really frightened me, and I had great concern as to what effect it would eventually have on my health. So I was very glad when after three months I was lucky enough to be transferred to another job, namely engine crankshaft grinding. This was wet grinding, and a continuous flow of water to which soda had been added to stop the rust was pumped on to the journals while being ground. This job was considered a picnic by comparison with the previous one.

Unemployment was high in those days, and if you complained too much about any job or the conditions prevailing on it you were labelled by the foreman as a trouble maker or a "stirrer", and told by him that "If you don't want the job I will have to do something about it". He could not sack you, but he could strongly recommend it. What a blessing it is that today's work force does not have to tolerate these conditions. However, during 1930 a programme was embarked on to install an extractor fan and ancillary equipment to every grinding machine on the premises. In due course this plan was completed, and subsequently every new grinding machine installed was duly fitted straight away with this very necessary feature.

Unfenced flat leather belting running from the overhead countershaft down to the lathe pulley was general in the old days. Only the train of gears at the end of lathe beds, required for screw cutting, was guarded. The standard of safety was exceedingly poor, and accidents occurred far too frequently. Happily nowadays the Executive, in common with all large industrial concerns, employs a Safety Officer, solely concerned with safety, and so far as practable he endeavours to prevent even occasional careless risks being taken. Protective clothing, gloves, clogs, goggles, etc., are now issued to all employees who are in any way subjected to hazard in the execution of their duties.

In the early days the floor gangways throughout the factory were marked by cast aluminium blocks. These measured 8ins x 3ins, and were embossed with the letters L.G.O.C., and a row of them was cemented into the concrete floor in the appropriate positions. All this has now been replaced by the modern system of yellow lines and signs, which are kept well painted and can be easily removed and repositioned should the need arise. The floors and gangways in the old days often became covered with a layer of hard greasy muck, a mixture of oil and dirt that was well trodden in. This could often be very slippery and dangerous, until a man was sent round with a scraper on the end of a broomstick to scrape it off. Today, however, the floors are kept much cleaner and are never allowed to get into this state.

The first-aid squads

Any accidents which occurred in the works were first dealt with by members of the First-Aid emergency squads. These were composed of five or six qualified St. John Ambulance men under the control of a leader. In order to retain their position as members of a squad the management very rightly insisted that all members should be re-examined in First-Aid knowledge and proved proficient at least once in every two years. In the engineers section we had three such squads, assigned to separate control points, where the necessary equipment was kept. When an accident occurred, in order to summon a squad it was necessary to sound the alarm. This was a siren, operated by the compressed-air system, and each of the three sirens had a distinct and different pitch, so that only the squad whose siren had sounded would be alerted and attend the accident. The squad members would immediately go to their respective control box, collect the equipment, and proceed to the scene of the accident. Here, under the direction of their leader, they would render First Aid, and either convey the patient, on a stretcher if necessary, or accompany him to the works surgery, where he would be handed over to the sister-in-charge. For this very necessary service, recognised officially by the management, the squad members were paid a sum of one shilling per week.

Many years later, in 1952, I was indeed honoured and privileged myself to be asked if I would agree to be appointed to the squad which served the shop where I was then working. I gladly accepted, and was duly appointed. To my mind the First-Aid squad in a factory such as Chiswick are a very necessary and important team of men. During the war years, when women were recruited into jobs in the factory, and in sections where a number of them were employed, a woman First-Aider was introduced into the squad which served that particular location. These squads could be called upon to deal with anything from a badly-cut finger to a major disaster. Following an accident there is often confusion, and sometimes evidence of panic, and so on arrival the squad leader accepts responsibility. The wrong diagnosis and subsequent action might cause immeasurably serious consequences, or even a fatality, and I was always very conscious of this basic fact.

For some unknown reason, in my experience the accidents always seemed to occur in cycles. There would be quite a long period free of accidents, and then two or three would happen very close together. One which I was called on to attend very soon after being appointed to the squad, and which I shall never forget, occurred quite near to where I was working. A blacksmith and his mate, called a striker or hammerman, were working over an anvil when a piece of metal flew from the job they were working on and struck the blacksmith in the wrist, severing the ulna artery. I had never previously seen arterial bleeding, and it was quite a frightening experience. However, we managed to stop the bleeding with a pad and pressure, and quickly transported him to the first-aid surgery, where we told the Sister what action we had taken. She promptly gave her instructions — Acton Hospital post-haste.

When I was transferred to the Metal Shop I was appointed as squad leader in the First-Aid team operating there. This squad covered the metal shop and saw mill, which to my mind were the two most accident-prone shops in the whole factory, for in the former there were heavy presses and guillotines, and in the latter there were high-speed timber saws and cutters on which it is very difficult and well-nigh impossible to give 100 per cent guard protection to the operator. However, I am pleased to say that our squad managed to cope with all the incidents that occurred here.

The compensation paid to employees injured while working in the service of the company, and thereby unable to continue their normal work, was in those days extremely poor and totally inadequate. It was known as Workmen's Compensation, and the rates paid were fixed by law under the Workmen's Compensation Act.

They were paid only while the employee was unable to work, and then only if the liability for the accident was acknowledged by the company. The rates paid were 15 shillings for the first week of incapacity, and 30 shillings for each succeeding week, and the ordinary Sickness Benefit could not be claimed at the same time. It is easy to imagine the distress and anxiety that was caused.

I subsequently had experience of this myself when working in the General Machine Shop in May 1933. I suffered the misfortune of having a very heavy driving shaft fall on my foot, it being part of my work to handle these shafts prior to machining them. I reported the accident immediately to the Sister-in-charge in the First-Aid surgery. She duly treated it, and recorded it in the Accident Book, and I signed this as a true record of how the incident had occurred. It had happened shortly after lunchtime, and I now returned to work, but my foot gradually started to swell. In order to keep on working, I cut my working shoe down the centre of the upper, so as to allow more room for my swollen foot. I should mention that in order to save wear-and-tear on our outdoor shoes it was the general custom in the machine shop to change into an old pair while working, because swarf, oil, and grease would quickly ruin any good shoes.

I then continued at work until the normal finishing time, when I hobbled down the drive and on to the tramcar to take me home. After spending a sleepless night my whole foot became so black and so swollen that it was impossible to walk on it, and to go to work was out of the question. My heart sank. Here was I, only just married less than a year, buying a new house on a mortgage which I could scarcely afford, and now to be deprived from working and earning the very necessary money. We called my doctor, but he simply said that I would have to rest my foot, and a medical certificate to this effect was duly sent to the works. After a fortnight at home I was at least very pleased, as the reader can well imagine, when he signed me off as fit for work. But could I go to work? Oh, No! I was asked to attend the L.G.O.C. Medical Centre at Lambeth North Station so as to satisfy the company's own doctor.

Here I was subjected to a gruelling and humiliating interview, and cross-examined as if I had committed a crime. I shall never forget how, with just one glance at my foot, the board's doctor said "Come and see me again in a fortnight". On hearing this remark I was really worried, and had to bury my pride, which was very hard for me, and appeal to him to allow me to go back to work as my own doctor had said I could. He very grudgingly agreed. Nowadays with paid sickness benefits, fair and reasonable compensation, and, what is even more important, a sympathetic, dignified, and human approach to all these problems it behoves employees to ponder, and to consider how lucky they are with the conditions now prevailing.

Machines and management

Right from the opening of Chiswick Works all the machines in the factory were painted a dark reddish-brown colour. This was a nondescript and sombre shade, and very depressing. Even when new machines were installed they were immediately repainted in this awful colour. However, in the late 1920s a project was started to paint all the machines in different colours. We in the workshops had the idea that this was done so as to relieve eye strain; whether or not this really was so I do not know, but certainly the various colours did brighten up the shops. The colours used ranged from dark blue to very light blue, and various shades of green and red were also adopted. However, after using this colour range for many months a standard colour was later adopted for all the machines, and tool cupboards as well. This was officially described as "hedge sparrow blue", though to most people it appears to be a light green. It must have been deemed to be satisfactory, because it is still in standard use today after more than 40 years.

In the original layout of the factory all the machines required to undertake the work of any one section were located in that particular section. Thus, for example, the worm grinder was in the Differential Section adjacent to the assembly benches for differentials, and likewise the valve, crankshaft, and camshaft grinders were located on the Engine Section. The power required to drive all the machines on any one section was supplied by one single electric motor. This drove the overhead mainshaft, which in turn drove the various countershafts which were mounted on gantries above the machines. This overhead shafting, and the gantries incorporating the countershafts, were supported principally on the main roof stanchions, and thus caused a terrific amount of vibration to the structure of the building. Under this system if the main belt driving from the motor should break the whole battery of machines was put out of action. Likewise, if only one machine should require to be operated the whole line of shafting would have to be driven. This was a very similar arrangement to that which had been common practice previously at the various depots and garages.

The policy prevailing in those days was for all chassis and mechanical units to be stripped down "to the bone", and for worn parts to be either turned or ground down to one of a predetermined range of standards, or, if this were not possible, then scrapped. Arising from the introduction of mass-production methods, which were inaugurated with the opening of Chiswick Works, it became essential that various standard dimensions for parts were maintained and that interchangeability was possible. Previously, when the overhaul work had been undertaken at the various depots, nobody was in the least concerned with standards. Worn shafts and bearings were turned down to a "clean-up" size, and bushes or bearings were made to suit. Oft-times unorthodox methods and procedures were followed, which, while achieving the ultimate result of getting a vehicle speedily back into service, could not be applied in a factory employing mass-production methods.

The accepted principle operating today regarding information which the management wishes to impart to its employees is that it is conveyed either by written material under cover in a sealed envelope addressed to each employee, or else through the medium of the notice board. But this procedure was not in being in the 1920s. In the first instance, to receive a personally addressed envelope was the ever-present dread of the workshop staff, likened unto the Sword of Damocles, for it could mean only thing; "Discharged, due to redundancy, services no long required", and as regards the latter there were no notice boards. This threat of discharge was ever-present and very real, and to understand the full significance of this it must be realised that only one hour's notice, by either party, was necessary to terminate one's employment.

It was the policy prevailing in industry generally to keep employees in the dark regarding future policies or ventures. In these circumstances it is very easy to imagine that rumour was rife. Information reached the staff mostly through the medium of the grape vine. Some of the news was grossly exaggerated, but was usually founded on basic fact often obtained through leakage in administration. It was understandable how this flourished, when information backed by authority was simply unobtainable. Such few notices as were issued, and these were certainly an event, were posted on the time-recording clocks. It seems stupid nowadays to have to relate, that changes in supervision or administration were not even recorded on a displayed notice; and as regards an introduction to the individual from whom you were to take instructions, this simply was not done. Vacancies occurring for supervisors or staff positions were never advertised, they were simply filled. How? I leave the reader to judge.

Since time immemorial, staff engaged in engineering workshops in the length and breadth of the country had always at some time or other performed an "irregular" job for themselves. Chiswick was no exception to this custom, and one would occasionally find the "foreign", "homer", or "private" job being undertaken.

By 1923-4, the next step forward in LGOC bus design was well under way, with the entry into service of the NS class of vehicles. The chassis had been designed for a covered top body and had a much lower frame level than the S but at first the Metropolitan Police would not sanction this, so the majority of the class entered service as open-toppers, though many were to be given top covers later.

Learning to drive on a B-type

All apprentices had to spend some time working on the chassis rectification section, which was where the final adjustments were made and all the odd little tasks completed that had got overlooked on the conveyor-belt assembly-line. This was quite a popular location to work for the vast majority of the lads, because the foreman in his wisdom would allow apprentices under the instruction of a tester to drive the various chassis around the works test track. This, I must say, was very desirable, and could well be very beneficial at some later date both to the company and the apprentice alike. Actually while I was there we were only allowed to drive the B-type chassis. There were plenty of the new K-type vehicles now coming through for their first major overhaul, but the foreman would not let us lads practise driving on them. The formalities for anybody driving were very few in the early days, but many years afterwards every person driving any vehicle of any type whatsoever on the works roads had to hold a current works driving permit, for whatever type of vehicle he was driving. These permits were issued only when the applicant had passed a driving efficiency test in the works driving school for that particular type of vehicle or vehicles.

I benefited myself from this early training in driving bus chassis around the works, although obviously an open chassis on a private track is very different from a complete bus where you are surrounded by bodywork and other traffic. Also those early B-type buses were vastly different from today's modern buses, being roughly only two-thirds of the length, half the seating capacity, and not much more than one-third of the unladen weight. If we compare the standard B-type with today's standard DMS-type Daimler double-deckers we find the overall length has increased from 22ft-6½in (which even includes a projecting starting handle!) to 30ft-10in (and a single-decker can be up to 33 or 36 feet long). The seating capacity has increased from 34 to 68, and the extra standing passengers from 5 to 21, whilst the unladen weight has grown from 3½ tons to 9.1/3 tons. The width has increased from 6ft-10in to 8ft-2½in, and the wheelbase from 12ft-10½in to 16ft-3in, whilst the overall height has jumped from 12ft-5in with an open top deck to 14ft-6in with a covered top.

What was the B-type chassis like? Instead of being made of steel channel like all modern vehicles are, the main frame was made of timber (ash), with steel flitch plates on each side, and was straight and high like a lorry's frame instead of being low down for most of its length and raised higher only over the two axles. As mentioned previously, the engine was a four-cylinder petrol type, with cylinders cast in pairs, and magneto ignition. Most of the B-types had a 30 hp engine with cylinders of 110mm bore and 140mm stroke, though some of the later ones had a 40 hp engine of 115mm x 140mm, as compared with the 170 hp of the "DMS" Gardner oil engine. The smaller B engine was used with a leather-faced cone clutch and three-speed gearbox, with a worm-driven back axle of 7.1/3 to 1 ratio, but the larger engine was used with a Ferodo-type single-plate clutch. The gearbox used sprocket wheels with chains engaged by dog clutches, instead of the more usual movable spur gears, and this gave a quiet-running system which could withstand a lot of rough treatment by careless drivers without damaging the box. Suspension was by ordinary leaf springs. Brakes were provided only on the rear wheels, with internally expanding shoes in a conventional drum, but there was also a so-called transmission brake. There was a clutch stop on the transmission shaft, or cardan shaft. When you pressed the clutch pedal and disengaged the clutch it then brought in the stop against a projecting circular collar around the shaft, to stop the shaft more quickly.

The bodywork was very typical of almost all other types, provincial as well as London, made during about 1905 to 1920, and seated 16 inside on longitudinal cushioned seats plus 18 outside (ie. on the top deck) on transverse wood-slatted seats. The lower saloon was illuminated at first by acetylene lighting, with a calcium carbide generator fitted under the driver's seat, but electric lighting was later introduced, with a dynamo driven off the engine flywheel. The General built all its own B-type bodywork in its own factories at North Road, Olaf Street, and Seagrave Road, except for 127 second-hand bodies transferred from older Milnes-Daimlers and Straker-Squires in 1910-11, and also for some new post-war replacements built in about 1919-20 by Hurst Nelson and Christopher Dodson. A total of at least 6889 B-type chassis were made by the A.E.C. at Walthamstow during 1910 to 1918, but less than half of these went to the General. Almost all those made during 1914-18 went direct to the Army as lorries, and a handful went to provincial bus operators, leaving only 3165 to be actually owned by the General. This figure includes 193 single-deckers, 41 vans and lorries, and 12 charabancs, of which in each of the three groups roughly about half had started life as such and roughly about half were subsequent conversions from double-deckers. A substantial part of the General's fleet of 3165 was engaged on War work in 1914-18; thus 1319 buses went overseas for troop transport, 1223 were held in reserve for emergencies, 233 were on London defence work, and 52 were on special air-raid service. Unfortunately a total of 465 were lost, mostly in France and Belgium, and did not return to the General after the war.

But I must return to Chiswick, and to my own apprenticeship. By the time I learnt to drive the quantities of B types were already

The B was still a familiar type of vehicle on London's streets. B770 is seen in early 'twenties livery. There were several minor body design variations on this model — note the slightly arched window outlines on this example.

The compact dimensions of the B-type are emphasised in this front view.

These six B-type chassis had been built with van bodies in 1911, for use by the ticket department. The photograph was taken around 1922, probably in the then new ticket building in Chiswick Works. From left to right, the vehicles are B703, 730, 1397, 1533, 724 and 1021.

dwindling, being augmented in 1919 to 1923 by 1062 new K-type and 902 of the further improved S-type chassis, which in due course allowed the last of the Bs to be scrapped or sold by the end of 1926, so that we did not overhaul many of them after 1925. I suppose my own driving instruction in the chassis rectification section must have been in about 1923.

To me at the start it seemed such a huge chassis to have under my own control. I found the clutch was the worst feature, for me at least, to get accustomed to. It was of the cone type, and immediately after overhaul before it had been 'run in' it could well be extremely fierce. The gate-type gear change was quite simple and positive in action, and provided the clutch was correct the gear change gave no trouble. The hand and foot brakes operated on the rear wheels only, as described above, and gave no problems to me as a learner. Although only a beginner I found it was very easy to drive a bus chassis, at any rate on the sheltered works test track, even though operation of all the controls was purely mechanical, and there was no power assistance in those days from compressed air or vacuum or hydraulic fluid, also no "self-changing" gearboxes.

However, I am not suggesting for one moment that for our bus drivers to control in public service, in all types of weather, a B-type chassis with the additional load of the bodywork and 34 passengers, was an easy way of earning a living, not even in those far-off days of nearly 60 years ago. Traffic in central London in those days was very dense, just as dense and congested as it is today, but the big difference was that most of it was still horse-drawn. Not only was the bus out in all weathers, but the driver also was. Above his head he had a little canopy projecting from the upper deck, to keep some of the rain off, but an enclosed cab was right out of the question. Glass windscreens were totally prohibited by the Metropolitan Police right up until 1931, more than a year after the arrival of the ST and LT types, with their improved standard of comfort for passengers. When first built from 1910 the B-type did have a small side window but in 1913 the Police even objected to this, on the grounds that it prevented the driver from giving proper hand signals, and so the General had to remove all of them and London bus cabs continued to have had no doors until the arrival of the 2RT2 type in 1939 and the Bristols, Guys and Daimlers of World War 2. On the B, and all other types until 1931, when it was wet or cold the driver had an extremely heavy overcoat, and goggles, also a canvas sheet which was fixed to the dashboard in front of him and which he tied up to the cab roof with rope after he himself had got in. In some cases this came very nearly up to his eye level, and was probably a very good substitute for a glass screen in the days before unbreakable or toughened glass became available.

Until mid-1931, the B-type buses had small side windows at each side of the driver, giving some slight degree of protection from the weather. However, the Metropolitan Police then objected on the grounds that the right-hand one prevented the driver from giving adequate hand signals. So all new B-type buses from that date had to be built without the windows and earlier ones were supposed to have them removed. However, the body on B2424 (above) had evidently not been modified when war broke out and as this bus was B2424 (above) taken by the Army in the 1914-18 War and, in this case, returned, it appeared in the simplified post-war livery with the side windows. B2578 (below) was one of the first to be built without the windows.

I never had the opportunity to drive a bus in public service myself, because of course the operating department was completely separate from the overhaul works. But the drivers' training school was nevertheless situated inside the grounds of Chiswick Works, and a very thorough and comprehensive course of instruction was given. Back in the twenties the candidates had to be between the ages of 26 and 35 (although nowadays the range is much wider), and to have already had at least two years' practical experience of driving heavy motor vehicles. They had to be not less than 5 ft. 7 ins. nor more than 5 ft. 10 ins. in height, because the standardised positions of the controls would have prevented a very short or a very tall man from properly reaching the pedals, and in those days seat heights were not adjustable. There was an extremely strict medical examination, and the course of training lasted up to a maximum of 34 days (though frequently less), during which a subsistence allowance of only 4 shillings per day was paid. But the original selection, plus the thoroughness of the instruction, was so good that only one candidate in fifty failed to pass the final test. New conductors also were trained in this same school. They had to be between 24 and 34 years old, between 5 ft. 6 ins. and 5 ft. 9 ins. high, physically very fit, and good at mental arithmetic. Their training course lasted a maximum of 14 days, and included fares and tickets, timetables and routes, the geography of London, the police and legal regulations, Safety First, the care of children and old people, and how to deal with obstreperous passengers.

In my own case, my instruction on the works test track on newly-overhauled B-type bus chassis was the only driving tuition I ever received. But it was to herald the start of many years of driving motor-cycles of various classes, also very many types of motor cars of varied manufacture and vintage. Soon after I was eighteen years of age (the minimum) I applied for a full public all-groups driving licence from the Essex County Council. No driving tests had to be taken in those days; one merely filled in the appropriate application form and paid a five-shilling fee, and the Licence was granted more-or-less automatically. This fee, by the way, renewable annually, remained the same for well over half a century.

The vast majority of the apprentices at Chiswick in my time obtained driving licences as soon as they were old enough, the great idea being to be able to drive any vehicle when an opportunity occurred. At this time motor vehicle insurance of any classification, although recognised as being desirable, was not compulsory by law, but the need to hold a current driving licence was enforced very strictly. In the early twenties if you were able to afford to run a car you were considered to be "somebody". Even solo motor cycles, and motor-cycle combinations, were not very numerous on the roads, hence the opportunity to drive was fairly infrequent.

So the engineering apprentice engaged on "bus work" was in a rather better position than most other young men. Often the owners of vehicles whose knowledge of maintenance or of running repairs was very limited would turn to an indentured youth for help or guidance. I must say that in my own particular case I was dead keen to offer any such assistance if I possibly could. On one occasion in the mid-twenties I was indeed fortunate to be asked if I would be prepared to drive a brand new car from the local show-rooms, and then to teach the owner how to drive his new vehicle. I was very pleased to do this, and as a result I had the pleasant experience of very many drives around the countryside in the days when very few other people were able to do so.

In those days street parking of motor vehicles of any description at all, including buses, during the hours of darkness was strictly prohibited without lights. No pavement parking was allowed under any circumstances at any time, and an unlicensed vehicle on the road was unheard of. It appalls me nowadays when I see would-be amateur mechanics lying underneath their cars, in the public road-way, usually outside their own home, and minus wheels, the weight of the entire vehicle being supported on a pile of bricks under perilous conditions. And nearly every road has a string of cars parked on both sides at night, without any lighting at all, very hazardous both for them and for the other cars that are trying to travel along. Back in the twenties our buses could be, and certainly were, stopped by the police for being in a dirty condition, and the private motorist was frequently "pulled up" for having a dirty number plate. Nowadays it appears to me that you can get away with driving most vehicles (although not a bus) without displaying a current licence disc; and regarding dirt it just seems nowadays to be of no consequence at all.

4. The Mid-'twenties

Indentured apprentices

Before the opening of Chiswick Works, when the overhaul work of the company's bus fleet was undertaken at the various garages, as outlined earlier in this book, any would-be apprentices were recommended by the Garage Engineers. It was then essential that these lads were already employed by the Company, had proved their worth, and in the opinion of the engineer were suitable and acceptable to become indentured. Following this recommendation, if approved by the Chief Engineer, the applicant would attend, at Electric Railway House, Westminster to be interviewed by the chief engineer.

Today the method of recruitment and selection of apprentices has vastly changed. The lads are recruited from school-leavers only, in the 15½ to 16½ age group. Many such written applications are received from would-be apprentices each week, and from these applications personal interviews are granted with the Apprentice Supervisor, who selects a proportion to sit for a written examination in mathematics. Governed by the result of this examination, and other circumstances deserving merit, the most promising lads are then requested to attend a selection panel. This panel is at officer level, and sits four times a year, and selects the successful candidates according to the number of vacancies. These lads are engaged when they are 16 years of age or over, not earlier, and after serving a probationary period of three months they are indentured to the L.T.E. Note that all the young men working in the factory today are apprentices, and the employment of young men as messengers, etc., classified as "improvers", has now ceased.

In the early Chiswick days the apprentices were moved from section to section on a rota system previously prepared by the Labour Superintendent (now designated the Senior Personnel Assistant), which allowed a specified period to be spent on each location. Once transferred to a section it depended largely on the initiative and enthusiasm of the apprentice as to whether his movement was restricted to a few tasks, or he received a fully comprehensive training on all the work on which this section was engaged. Some of the supervisors were not unduly concerned as to whether or not the apprentices left their sections with more knowledge and experience than when they came. These worthy gentlemen were too output conscious, and the apprentices were often rather unfairly used to boost output on repetition jobs, although in fairness to most of the supervisors, if an apprentice was really keen they were usually very co-operative and ready to help him.

The Company allowed the apprentices up to one hour's leave for three afternoons per week to enable them to attend evening school for further education. In those days this was considered a really wonderful and generous concession, and every apprentice was expected to avail himself of this privilege, which in fact the majority did. Evening school work was hard and tiring when one had already completed a hard day in the factory, but it paid dividends both to the company and to the apprentice, and I am sure both parties reaped some benefit from the knowledge so gained.

To be an indentured apprentice to a company of the size of the L.G.O.C. was considered to be an achievement, and a lad was deemed to be lucky to be so engaged. The policy of the Board of Directors of this company and its officers was indeed kindly disposed to these lads, and I am quite sure these people desired the terms of the Indenture to be carried out to the full. Unfortunately some of the administrators looked on the lads as a disagreeable nuisance, but if an apprentice could not get the training and experience that he was entitled to, he could complain to his Section Engineer. If then still dissatisfied, he could go direct to the Chief Engineer of the Company with his complaint. I have no hesitation in saying that if the terms of the indenture were not being observed it was the apprentice's own fault.

I can relate here a classic but typical personal example of the lack of interest and concern for the welfare of apprentices on the part of foremen and supervisors. When I had completed the first two years of my apprenticeship my father asked me at home if I had yet received my additional 1d an hour merit rise which was fully outlined in Paragraph 5 of the indenture, and about which the Chief Engineer had told him when we had both signed the indenture to start the apprenticeship. I told my father "No", and he replied "You had better remind your foreman about it". At this time I was working on a lathe on the engine section at Chiswick, and I promptly asked my foreman if he had any information about my merit rise, which was now overdue. He replied, "Leave it to me", but, characteristic of the general apathy at this time of the supervisors to staff matters, nothing more happened, and this gentleman could not have cared less. On learning this my father told me to ask my foreman again, and this time I was told not to worry him, but to leave it to him. Once again nothing happened, and my father became concerned that the terms of the indenture were not being carried out, owing to the attitude of some petty official of the company.

He certainly was not going to let me be a victim of this lack of interest and consideration in what he believed was my right, so he wrote a letter direct to the Chief Engineer requesting an explanation. Two days later I was busy working at my lathe when a clerk from the office came up to me, and told me I was to go with him to the Works Engineer's office straight away. When I saw this worthy gentleman he said, "I have a letter here that your father has written to the Chief Engineer, but why on earth didn't you come and see me?" He was extremely annoyed, but I told him my father did not ask my permission before he wrote any letters, and in any case I had asked my foreman on two separate occasions about the subject with no response. This was a very typical example of the attitude of the lower management to the works staff, and I was glad that in this instance it had been exploded and brought to the notice of the top management. I am glad to say that I received my merit rise on the very next pay-day, together with full back-pay from the time I had completed two years of my apprenticeship.

The apprenticeship itself was very comprehensive, and there was plenty of scope for learning, and some of the extremely capable and clever employees were willing to instruct the eager and willing-to-learn lads. There was no Apprentices Supervisor as there is at the present time, when this gentleman's sole duties today are concerned with the problems and welfare of the apprentices. In my day when an apprentice had completed the terms of the agreement as per the indenture, usually on his 21st birthday, he was handed the document duly signed by the Chief Engineer or the Engineer-in-Charge.

Time served—qualified engineer

I completed my apprenticeship with the L.G.O.C. on the 7th of April, 1925, which was my 21st birthday. During the last three months of my time I had been employed in the drawing office, and had found the work very absorbing and interesting. It was quite a small office, with two draughtsmen engaged on body work and four on mechanical work, assisted by two apprentices, with a young lad to do the filing of the drawings and to operate the drawing printing machine. There were also three men on the specification section, one on body work, and two on chassis work. I evidently made my mark in the office, for the chief said he would very much like to keep me on his staff, but, alas, this was not to be.

On the very day I had completed my time I was summoned to the labour bureau, from where the clerk escorted me to the office of the works engineer. I was lucky insofar as that I was wearing a smart suit, but had I been working in the works at the time I should have been taken to the chief all grubby and dirty. I always felt that on an occasion such as this to be hauled before the chief straight from the workshop clad in dirty overalls tended to humiliate you and make you feel very conscious of your relative position. This, I am sorry to say, was sometimes done deliberately in this era, as also was to keep the person waiting outside the office stewing as to what his fate could be. This was a very dodgy period in which to complete one's apprenticeship, and just prior to this several ex-apprentices had been discharged as no work could be found for them, so I was very apprehensive as to what was to be my fate!

The works engineer received me quite formally, and, with a word or two, handed me my indenture, on which had been added the following:- "We hereby certify that Henry Desmond Scanlan of 134 Halley Road, Forest Gate has served his full time under this Indenture. He has displayed the keenest application in all branches of work through which he has passed, and has given us the utmost satisfaction. (Signed) GEO. J. SHAVE, CHIEF ENGINEER. 8th April 1925."

He gave me the usual "old flannel" that was so characteristic of those days, saying that the company did not wish to lose the young men whom they had trained, regardless of the fact that some had recently been discharged, and he would be pleased to offer me a position in the engine section as a crankshaft grinder, a job which he understood I had done some months previously. This in fact was correct, but it certainly was not a job I would have chosen had I been given the chance. However, the employment position at that time in the whole country, including the engineering industry, was extremely bad, and I was thankful to accept this job, for it was better than doing nothing.

At that time Wednesday was the beginning of the factory week, and so on the following Wednesday I started work as a skilled man on a crankshaft grinding machine in the Engine Section. There were eight such machines of the type I was assigned to, all of Landis manufacture from America. Four of these were equipped to cater for the grinding of the big-end journals, and the other four for machining the main journals. The crankshafts had to be lifted by hand into and out of the machines, which was not a very easy job. A fine degree of accuracy and finish was demanded for the grinding, and a specified output was required from each machine. At the onset it was quite a reasonable job, but the flow of engine work would ebb every now and again, probably after a period of from six to nine months, and men would be discharged, followed very soon afterwards by a cry from the management that "More output was required". This procedure was sickening, but, alas, all too common in industry at this time. To be fair to the company, though, it was a limited-liability private-enterprise organisation, run for profit for the benefit of the shareholders, and the human problems associated with the business were only too often secondary considerations in the planning and operation of any projects which the management envisaged.

Bus overhaul in the 'twenties

With the opening of Chiswick Works the time taken for the complete overhaul of a bus was reduced from sixteen days to eight days, and after a few years it was reduced to only two days. A booklet published by the L.G.O.C. in 1929 gave a detailed technical description of the whole place, and in this chapter I would like to summarize some of its main contents. The whole site occupies an area of 32 acres, of which the buildings (at the stage of development reached by 1929) covered some ten acres, if we include all the office blocks and ancillary buildings. The one main factory building, all under the one roof, was rectangular in plan, covering seven acres, being 855 feet long by 350 feet wide. The walls were of concrete slabs, and the roof of steel-truss type, covered with asbestos tiles. The Plenum system of heating and ventilating was installed, in which hot air at 60° F was distributed throughout the building in winter whilst in summer cool air was circulated. Precaution against fire was provided by an automatic sprinkler system. Electric current for power and lighting was obtained from the Underground Railway power station at Lots Road, Chelsea, being taken from the Acton sub-station of the adjacent District Railway and transformed in a sub-station on the northern edge of the Chiswick site. The Test Hill, or "Dip", which I mentioned earlier, is 300 yards long, with a gradient of 1-in-15.

The original boiler house (at the northern end of the original Stores section in the middle of the main factory) contained four "Economic" boilers for supplying steam to various washing and cleaning appliances and also to the works heating system. Adjoining the boiler house was a rag-washing and oil-reclaiming plant, with three turbine oil extractors, one water extractor, three washing machines, and two drying machines, in which all dirty rags collected from the factory and from the garages were dealt with. The oil was first extracted, and used as fuel in an oil-burning boiler, and the rags were thus reclaimed. All round the factory overhead runways were installed for the rapid transit of materials, and many other labour-saving devices were in force. The latest designs of pneumatic tools and machines were installed, and the plant throughout both the body and the chassis sections of the factory was of the most modern type. All the machines, of whatever kind, necessary for the repair of a particular component, were collected into groups, rather than segregating the various types of machines and having a special section for each type.

On arriving at Chiswick for overhaul a bus was driven straight to the Dismounting Section, where route boards, lifeguards, wings, lamps, cushions, and other fittings were removed. The body holding-down bolts were undone, and then the complete body was raised from the chassis by a hydraulic lift, and mounted onto a four-wheeled bogie on which it could be moved around the factory either by hand or by Fordson tractor. The body was taken first to the Stripping Shop, for the removal of all defective parts, and then passed into the Body Repair Shop. Here the bogie was attached to a ropeway moving at four inches per minute, which slowly carried the body along between five groups of craftsmen. The first group (and remember that we are dealing here still with largely open-top outside-staircase bodies) carried out all necessary repairs to the platform, fender, stairway, side panels, and staircase stringer. The second group dealt with top-deck seats, front and rear panels, side frames, and gutters. The third group worked on the roof joints and slats, front and rear boards and fittings, windows, lower-deck seats, and panels. The fourth group attended to the ventilators and traps, top-deck side-rail and ironwork, and bell fittings. Finally the fifth group repaired the lower-deck floor slats, apron rails, rear advertisement frame, switch box, and electric wire mouldings.

The repairs completed, the body was transferred, on the same moving ropeway, into the Paint Shop. Here the "flow" method was adopted for the sides of the body as far up as the waistline. One flow coat, from something rather like a garden watering-can, was equal to two coats put on by brush. One coat of grey and another of red was given, after the undercoat had been applied. The red paint imparted a semi-glossy surface, so that only one coat of varnish was necessary, and the fleet name and number transfers were put directly onto the red paint whilst it was still tacky. Surplus paint dripped off into V-shaped troughs standing on the shop floor. The bus roof and top deck were painted by hand, as also were the interior of the lower-deck and the pillars and certain other parts of the exterior. The paintwork having been lined out and the transfers fixed, the final coat of varnish was given. Bodies took about two working days of 8½ hours to pass through the Paint Shop, so as to allow time for the paint to dry. The full overhaul of a body took a lot longer than the overhaul of a chassis, and for that reason the same body never went back onto the same chassis (except for special non-standard types of bus), and with every new batch of buses that were purchased there were always a few spare bodies, in the ratio of 103 bodies to 100 chassis.

The painting finished, the body was lifted from its temporary bogie by another pneumatic hoist, a freshly overhauled chassis was driven underneath, the body was lowered and bolted down, and the re-assembled bus was taken to the Mounting Shop. Here it was fixed to another ropeway, travelling at fifteen inches per minute, and moved along whilst lifeguards, wings, lamps, cushions, route and advertisement boards, etc., all of which had meanwhile received

Metropolitan Police requirements for frequent overhaul of London buses justified volume production methods. Here an S-type body, having been removed from its chassis and mounted on a trolley, receives attention from five craftsmen.

(Below) Trimmers attend to seat cushions. This is a 1922 picture—note the B-type buses in the background.

(Below, right) The Destination Board shop, also photographed in 1922. The racks in the background contained place-name labels which were pasted on the boards before being varnished.

attention in the Mounting Shop, were replaced. Apart from testing, the bus was now ready for service.

The conveyor-belt system of bodywork overhaul described above was not in fact adopted until 1926. For the first five years at Chiswick the bodies were mounted on trestles and remained stationary, as they had also been in North Road, Olaf Street, and Seagrave Road. The complete overhaul of a body then took from four to seven days, but when the flow-line system was introduced the passing of a body through the Repair Shop (excluding painting) was reduced to only eight hours or so.

Now we turn to the chassis overhaul, which was the side that I personally was involved in. On removal of the body the chassis was taken at once to the Chassis Section, where any petrol remaining in the tank was drained off into an underground receptacle. The chassis was then attached to a ropeway moving along the eastern wall of the factory at one foot per minute, and was stripped of its components in this sequence:- petrol tank, radiator, engine, steering, brake rods, cardan (ie. propeller) shafts, gearbox, front axle and wheels, front springs, rear axle and wheels, and rear springs. On removal, all these components were passed through washing machines, of which there were two, one for the main frame and wheels and large items, the other for all the smaller items. Both consisted of sheet-iron tunnels with perforated longitudinal tubes, through which boiling soda-ash solution was pumped onto the components as they passed through the tunnel on a moving conveyor, and emerged at the other end, thoroughly cleaned, into the Viewing Section.

Here, everything was carefully inspected with special jigs and limit gauges. Any parts that were worn out or needed repair, such as ball races or gudgeon-pin bushes, were removed, and the condition of every unit was indicated by dabbing a blob of paint onto it. Thus green was put onto items that were fit for immediate re-use, blue meant that rectification was needed, and red implied scrap, and hence any chance of an unfit part being used again was eliminated.

The first item dealt with after viewing was the radiator. The top and bottom tanks were removed, defective tubes were taken out, and the tube centre was dipped in caustic soda and then immersed in a corrosive fluid to remove the scale. The tube centre was then rinsed and sent by conveyor to the bench where new tubes were fitted and soldered. The complete tube block was tested by air for leakage, and the aluminium tanks were treated with cleaning acid and refitted, the whole job of repairing the radiator taking only twenty minutes. Dashboards, bonnets, and petrol tanks were also repaired in this same shop, usually known as the coppersmiths.

Meanwhile in the Engine Section the whole engine had been dismantled and cleaned and its components put through the washing machine. With the aid of a die-casting machine the engine main

bearings and connecting-red bearings were relined at the rate of thirty per hour; thus one man could reline in a single day all the bearings for fifteen engines. Crankshafts were straightened, centred, and reground to one of nine standard diameters, and a tolerance of only .0005'' (half a thousandth of an inch) was allowed. A battery of eight Landis grinders, on one of which I worked for a long time, gave an output of 42 crankshafts daily, and the crankshafts were then fed into a special crankcase-bearing reamering and running-in machine which was designed by L.G.O.C. engineers.

This machine was constructed on the turret principle, and did four operations all at the one setting. The first operation was in removing one finished crankcase and locating another, ready for reamering, onto the jig. The second was the reamering of the bearings to a standard size. The third operation was the renewal of the main bearing caps and the fitting of the crankshaft, whilst the fourth was the running-in of the crankshaft. Four men were employed on this machine, and accomplished in only twenty minutes the work that formerly took one man two days when bearings were scraped by hand.

With its shells and shaft fitted the crankcase now began its three-hour journey along the assembly conveyor. This was a moving platform at bench height, about 120 feet long, travelling at 7½ inches per minute, thus giving 16 minutes for each of twelve stages of assembly. In the first six stages the connecting rods and pistons were fitted to the crankshaft. The engine was then removed from the conveyor, by a pneumatic hoist on an overhead runway, to the test stand nearby, where it was run in. If the test proved satisfactory the engine was replaced on the conveyor, where at succeeeding stages the carburettor, magneto, dynamo, etc., were fitted, but if any defects were revealed in the test the engine was transferred to an adjacent rectification bench so as not to disturb the regular flow of work.

The component parts of the engine had meanwhile been transferred by the material conveyor, which ran parallel to the assembly

(Opposite page, top) NS chassis being pulled along the assembly line at Chiswick. Here the wooden front mudguards and lifeguards are fitted. From September, 1924, new NS chassis were assembled at Chiswick instead of being supplied from the AEC factory at Walthamstow. This lasted until 1928 when manufacture for LGOC was resumed by AEC after the latter had moved to Southall. This photograph was taken about 1925.

(Right) Crankcase boring, reamering and running-in machine, constructed on the same principle as a turret lathe, to perform a sequence of operations on each crankcase. This photograph was taken in 1922 and shows B-type crankcases.

(Far right) A 1926 photograph of the chassis assembly line showing the rear view of a reconditioned NS chassis being assembled.

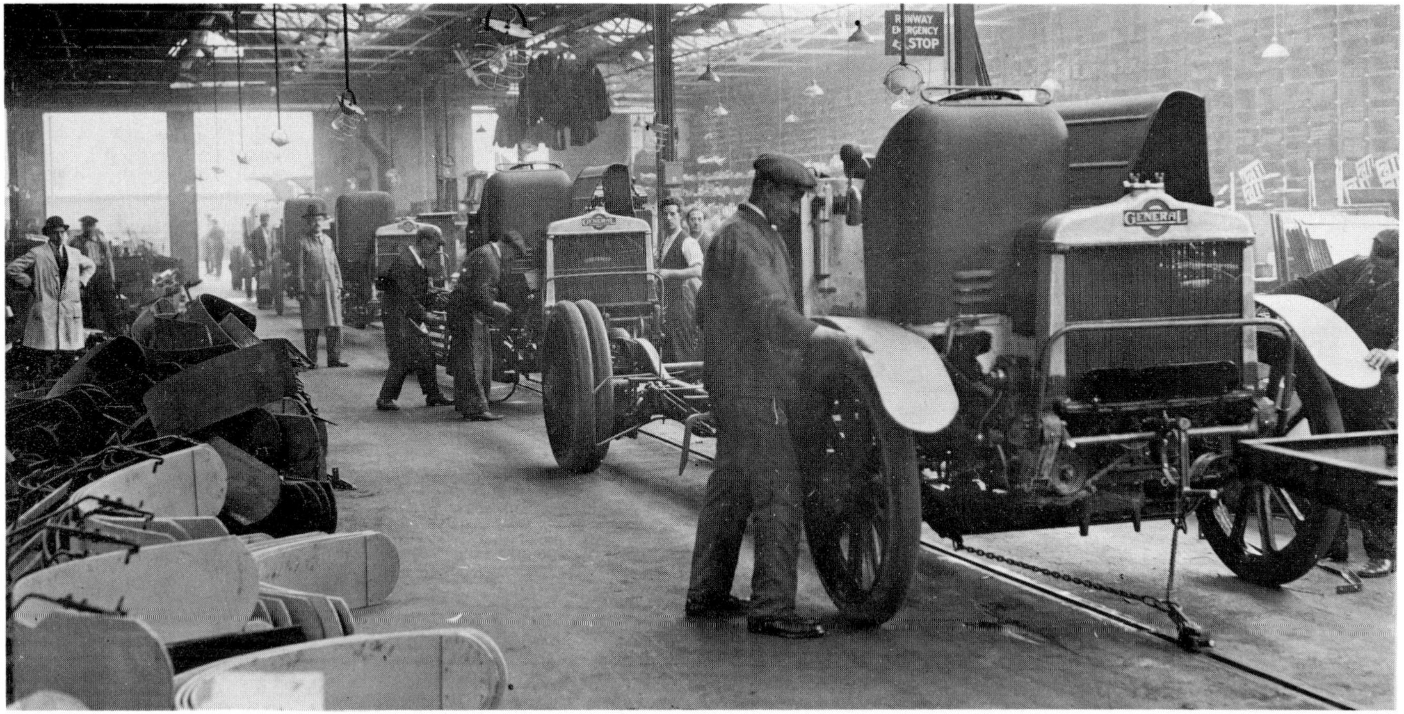
RUNWAY
EMERGENCY
STOP
GENERAL

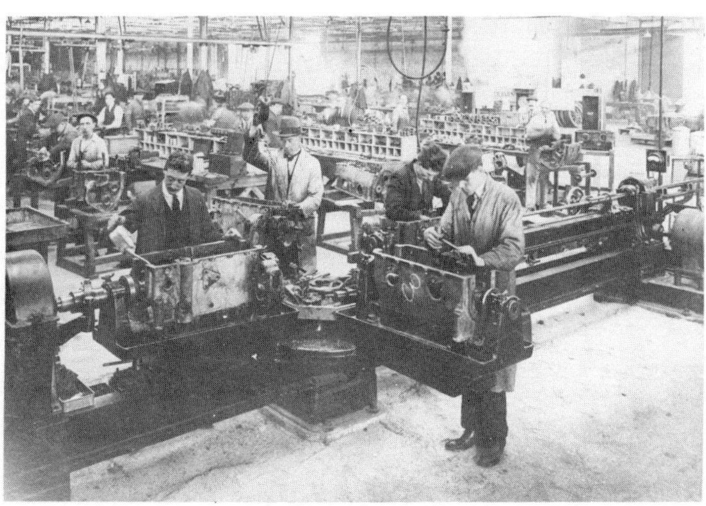

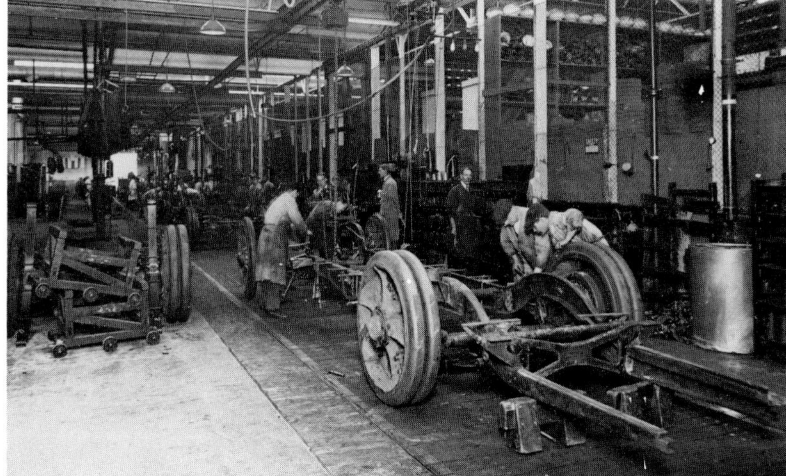

conveyor but about fifty feet away, and arrived in the sections where suitable machines rectified their faults. From here they were delivered to the particular stage of the assembly conveyor at which they were required for re-assembly. In 1918 at the garages the total time taken to overhaul an engine was about 85 hours. In 1919 with the adoption of the squad system this was reduced to 51 hours, and now at Chiswick it was only about 29 hours. In those days bus engines needed overhauling far more frequently than the modern diesel does. As I said, we reground 42 crankshafts a day, which was roughly about 11,000 per year, or roughly about 2½ times as many as the chassis overhauled per year. Engines (and other units too) were constantly being removed at the garages and sent in to Chiswick by lorry for another major overhaul, indeed sometimes they came back after only six or eight weeks, although if this was found to be due to faulty workmanship somebody got into serious trouble. Engines also frequently came in for what was known as a "partial" overhaul, where they did not need to go through the full treatment.

By similar methods all the other units of the chassis were overhauled and re-assembled. There were separate sections, each with their own moving flow line, for dealing with clutches, steering, gearboxes, front axles, chassis frames, differentials, and rear axles. All these lines were parallel to the engine line, and located, in the sequence as quoted, northwards from the engine section and in the middle of the main building. The so-called detail section dealt with the oddments such as cardan shafts, brake levers, change-speed levers, etc. Wheels and springs were dealt with in a separate section in the far corner of the building. The tyres, and of course these were mostly still solid tyres in the nineteen-twenties, were removed from the wheels, the wheels were inspected and any cracks in the spokes were welded by oxy-acetylene or electric-arc welding, and new tyres were fitted. Brake drums were bored and fitted, and the output was 500 wheels per week. All springs were taken apart, and the leaves re-set, hardened, and tempered, using a reversing regenerative gas furnace, ten feet square, in which leaves were uniformly heated to 900° C. The re-assembled springs were passed through another furnace on a conveyor belt, and then lubricated with whale oil and tested for load camber.

All the various components having gradually moved through the factory from east to west, the chassis was now ready for re-erection. This was carried out on a moving platform 220 feet long by 8 feet wide, which travelled from north to south along the west side of the shop, at floor level and at a speed of 15½ inches per minute. Ranged along one side of this platform were all the units required for re-erection, each unit being located at the exact stage at which it was required. Firstly, the front and rear axles, complete with springs and wheels, were positioned on the track and the chassis

frame was then lowered from an overhead runway and bolted up. Next came the gearbox and the rear cardan shaft, followed by pedal and brake gear, steering gear, the complete engine and clutch, the radiator, dash-board, and then the bonnet. Each unit was erected by its own squad of men, who did this job only. The final operations were the coupling up of water pipes, the adjustment of brakes and engine control, oiling and lubrication, and the filling of the petrol tank. The engine was then started up mechanically by a device consisting of revolving drums at floor level. The gears were set at neutral, and as the drums rotated the back wheels of the bus the engine was started up by engaging top gear and releasing the clutch.

If the running of the engine in the shop thus met with approval the chassis was then handed over to the Testing Department, who drove it around the test track that skirted the factory, and up and down the test hill, and also over a stretch of prepared bad road. The pulling-power of the engine, also the brake efficiency, were submitted to severe tests, and any rectifications that were needed were made. The chassis was then handed over to the Licensing Department, who had to see that it was in every way up to the high standard of efficiency demanded by the Metropolitan Police. This done, the chassis was taken to the Body Section, where the overhauled body was remounted, and the complete bus was then sent to the Metropolitan Police for final passing out and relicensing.

To the modern reader accustomed to the modern mass-production manufacturing and assembly methods in the automobile industry today the foregoing description probably does not contain any great surprises. But fifty or more years ago, in the early days of Chiswick, it was certainly quite revolutionary. At that time the General had probably the largest motor-bus fleet in the whole world, and certainly nobody else had previously used the conveyor-belt system for bus overhaul on such a large scale. In 1929 the General had a fleet of 4380 buses, mostly of the famous K, S, and NS types, which worked a total of 173 million miles per year, giving an average mileage per bus of 39,500 between major overhauls. Chiswick was overhauling buses at the rate of one every half-hour, 17 every day, or about 92 every week, for which about 350 machine tools were installed in the Works. Including ancillary functions such as the administrative offices, the Purchasing Department, and the canteen staff, etc., a total of well over three thousand people were employed at Chiswick.

Although the description I have given in this chapter is a summary of one published by the Company in 1929, most of the facts quoted would be true from 1925 onwards, and some of them of course right back to 1921. They say that Rome was not built in a day, and, like all big things, Chiswick took several years to grow from nothing to its full size and full output. When it started, the fleet was mostly the

Newly overhauled S-type body and chassis are united in the Body Section at Chiswick Works. The chassis about to be reversed into position was S197, built by AEC at Walthamstow in 1921, about the same time as the body, number 5750, which was built at the LGOC body shops at North End Road before Chiswick Works was opened. Note that the painting of the chassis was only partially completed.

very small B-type bus, but all the equipment and conveyor-belts and machines and washing tunnels, etc., were large enough to be able to cope very well with the much larger LT and STL types of bus in later years. Various small additions to the premises were being made all the time, but it was not until 1938, when the fleet had grown to over 6000 buses and coaches, that a really substantial enlargement and re-arrangement of the facilities was started on. For my own part I am glad that I was fortunate enough to be in right at the very beginning in 1921 and see the whole thing grow.

Pay day—and income tax

Right from the beginning of the omnibus industry, the payment of wages due to bus workers in most undertakings was made on Fridays. In fact almost all industrial undertakings in other industries had paid their staffs on this day of the week. This was indeed the traditional national pay-day. The London General Omnibus Company was no exception, and wages were paid in cash on production of the appropriate clock card which had been received previously from the foreman. At the various garages before the opening of Chiswick Works, the wages were paid by the garage cashier from his office. The staff lined up outside, just prior to the lunch break, so as to take their turn and receive their dues at the office window.

With the opening of Chiswick Works the payment of wages was still from the cashier's office (the location of which still remains today in the same corner of the main office block), but it was not from the office as such. Instead it was from the outside windows, where long covered gangways had been constructed, and personnel had to line up as nearly as possible in the numerical order of their clock cards to receive their wages which had been previously prepared in an unsealed envelope. This arrangement continued for many years, until a number of portable booths were constructed (in the works), and wheeled in every Friday to various predetermined locations inside the factory. Staff thus received their wages closely adjacent to where they actually worked, instead of (in some cases) having to take a very long walk to the far end of the building. This system still operates today, and seems entirely satisfactory to both management and staff. The actual pay day was changed some years ago from Friday to Thursday, and then to Wednesday, which it remains at the present time, but on whatever day the wages are paid that day without a doubt is the favourite and most important day of the week.

Starting early in 1920 all the administrative staff of the L.G.O.C. were paid monthly and by cheque. In those so-called "good old days" it was considered by many people that to be paid monthly gave added status to one's employment, and if by chance you were able to have your cheque paid direct into your own banking account you were certainly considered to be somebody of note.

Before the Second World War the income tax payable by individuals was their own concern. How much they paid, and how, was entirely their own business, and their employers had no knowledge whatsoever of the amount of tax, if any, that they paid. All the employer did was to furnish the Inland Revenue Authorities with a note showing the total amount of earnings paid out during the income tax year. Until this time comparatively few workshop staff qualified to have to pay any income tax at all, because most of them did not receive high enough wages. It was usually only single men, or married men without any children, who were liable, as the tax allowances given to married men with young children gave them exemption.

In October 1940 a compulsory system of income-tax deduction was introduced, under which the board, in common with all other employers, was required by law to deduct from the salaries and wages of their employees the amounts of income tax advised to them by the Inland Revenue Authorities. Then in April 1944 the present and much more complicated "Pay-as-you-Earn" system of income tax deduction was introduced.

An overall for a shilling

During 1925, arising from a suggestion submitted by an employee in the workshop, the management made the necessary arrangements to make it possible for employees to purchase old secondhand white cotton coats for use as overalls in the works. In those days these white coats, when new, were used by road service drivers in the summer as an alternative to their heavy blue serge overcoats in the winter, but nowadays slate-coloured coats with blue cuffs and collars are issued instead of white, perhaps because they don't show the dirt so easily. Two of these white coats were issued free to all bus drivers, but after a predetermined period they were instructed to hand them in and were then issued with two new replacements.

The old coats were often still in quite good condition, and hence made excellent overalls, especially for those staff who had a reasonably clean job. The management set the extremely reasonable purchase price of one shilling per coat. This was later increased to one shilling and threepence, and then to one shilling and sixpence, but, even so, these white coats were still extremely good value for money, and served a very useful purpose. They made us all look more important, and cut across the old and almost universal tradition in the engineering industry whereby only the foreman wore a white coat, and all the rank-and-file wore a blue boiler suit, with the charge hand wearing a brown coat.

Sales of these ex- bus drivers' coats continued until the outbreak of war in 1939, but then consequent upon the introduction of clothing coupons in 1940 the facility was withdrawn, and has never been reintroduced. Nowadays, with the consent and co-operation of the management, a private concern operates a fortnightly laundry replacement service both for boiler suits and for overall coats for all the members of the staff who wish to avail themselves of this facility.

In 1925, the NS with covered top was at last approved by the Metropolitan Police, over two years after the first prototype was rejected by them. This photograph of NS 1734, the first of the four buses which entered service, was taken in August 1925—even then there was to be further delay until October before they carried fare-paying passengers on route 100 from Loughton to Elephant and Castle. The chassis were among those assembled by LGOC but the bodies were built to LGOC design by Short Bros. at Rochester. Approval soon followed for 200 more and the covered top henceforth became standard for London double-deckers.

LGOC's standard method of producing the covered top NS body was to build the top-deck on its own floor, complete even to glazing, and then attach it to the lower-deck pillars. Compare these Chiswick views of an open-top NS body being built in 1925 and a closed-top body being assembled, with the lower-deck sunk into a pit in the floor, taken in 1927.

The General Strike in May 1926 produced some strange scenes. Here eight B-type buses are seen in the main drive of Chiswick Works being prepared for service by volunteer crews and under armed guard. They all display the special destination board for the circular route 1 from Ealing Broadway via Kensington Church, Aldwych, Notting Hill Gate and back to Ealing Broadway.

The General Strike and its aftermath

As a result of the General Strike that occurred early in May 1926, all the principal transport undertakings throughout the length and breadth of this country were brought to an abrupt standstill, and the London General Omnibus Company's bus fleet was no exception. It is not proposed to enlarge on the merits for or against this nation-wide crisis; suffice it to mention that a very small percentage of the staff of the Company continued to work. The bitterness, resentment, and dissension caused through this calamity is best not pursued. But hereunder is a copy of a letter sent by Lord Ashfield, Chairman of the Underground group of Companies, to all members of the staff who, to use his own phrase, "remained loyal":-

55 Broadway, S.W.
8th May, 1926.

"To those employees who have remained loyal during this great crisis:

"I have issued a circular letter of thanks and appreciation addressed to those who are helping us to re-establish our services, and what I said in that letter applies particularly to you. But I feel that a special word of praise is due to you, and on behalf of the Board of Directors, our principal officers, and myself I extend to you our best thanks for your loyal and devoted service in this great crisis. We realize the special difficulties under which you are carrying on and we trust that it will be of some satisfaction to you to know how much we all appreciate what you are doing.

"Signed: — Ashfield".

Many of the volunteer bus drivers were car owners, and they parked their cars on the Chiswick sports ground before changing to the NS and older buses on the left. Car owner-ship in those days in itself implied comparative affluence, but several of the cars are relatively large models.

The bench seat at the front of B5088 accommodates a volunteer driver, a companion and a policeman on the emergency service 1 during the General Strike. Note the barbed wire round the bonnet.

Before closing on what the author would describe as the most regrettable industrial episode occurring in the whole of the period covered by this manuscript, hereunder is a copy of the certificate given subsequently to all "who remained loyal".

General Strike
1926
London General Omnibus Company Limited.

The Company wish to heartily thank   who acted as a **Volunteer** during the General Strike, 4th to 14th May, 1926, for his services in maintaining London traffic and thereby coming to the support of the Country in a serious crisis.

This colourful certificate, which measured 9½" x 8", was designed incorporating flags, crests, and the Company's monogram, and signed by the Chief Engineer and Operating Manager, and the Chairman. All employees who qualified for one of these certificates received it duly signed, sent by post to his home address. When agreement was reached and the strike declared off, the resumption of the normal routine of the works was commenced, and gradually the staff were informed by letter sent through the post when to report to their sections to start work. The start of the strike had been very definite, but the resumption of work was very gradual. It was many weeks before the works were more or less back to normal. It is interesting to note that to date this was the last serious industrial conflict which affected those engaged on bus work at Chiswick.

Until the General Strike of May 1926 the comradeship amongst the fellows working together on the Shop floor was extremely good, and all were friendly and helpful to each other. We had one or two typical "yes men", so servile that I think they would have licked the foreman's boots as well as trying to do his job for him if they thought that by so doing they would receive some recognition. But fortunately they did not get very far; the other men made sure of that.

Unfortunately, however, when the strike was over and normal working was finally resumed, owing to differences of action and opinion regarding this conflict the atmosphere had completely changed to give way to distrust, tension, and intense bitterness. Friendships between parties were severed, and the lunchtime card schools and many social events were broken up over this issue. One particular fellow whom I knew very well had committed what was in workshop code an unforgivable crime. To my mind he was indeed very foolish to push his luck and really cash in on his fellow workers, whatever their action or the principle which was involved. His work-mates swore to get him for his conduct, and he had to ask for police protection as feeling ran so high as to be really dangerous. However, luckily he was quickly transferred from the factory, though whether from his own choice or not I can only hazard a guess. It took very many years to return again to a nearly normal working atmosphere.

Re-grading and increased pay

However, for those of us engaged on bus work for the company the year finished on a very much brighter note. Hitherto all employees engaged in the workshops were graded in their various trades, i.e. blacksmith, fitter, coppersmith, coachmaker, painter, etc, and likewise the semi-skilled were similarly graded as unit adjuster, hammerman, bench hand, brush hand, etc. Each individual grade carried its own preset basic rate per hour worked. The company now envisaged a scheme to amend this system, which did in fact considerably improve the earning capacity of all employees.

Hereunder is the new 1926 regrading scheme, which needless to say was received with great satisfaction by all concerned:-

"THE LONDON GENERAL OMNIBUS COMPANY LIMITED.
"TO THE STAFF EMPLOYED AT CHISWICK WORKS.
"I desire to express on my own behalf and that of the Board of Directors and Officers of the Company, our satisfaction with the progress which has been made at Chiswick Works during the year. The efficiency and the output of the Works has increased and this result is due to the excellent team work which has followed from the improved relations between the Management and the staff. It is our intention that the staff shall share directly in the economies and benefits which accrue, and the following Notice sets out our present proposals. We hope that through further improvements it will be possible to still further advance the interests of all concerned.

"Commencing on the first full pay week in January 1927, increased wages will be paid and the various grades concerned brought under three groups as shown below.

"These increased wages include all allowances and will be paid on the understanding that the existing rates of pay and the conditions relating thereto shall be deemed to remain unaltered — the amounts over and above the existing rates being regarded as in the nature of a bonus and subject to revision in accordance with the measure of efficiency and output secured from time to time.

"It is expected that all grades concerned will be willing to perform such duties as may be required of them and which they are competent to perform.

"GROUP 1 — Craftsmen — existing rates plus bonus: 1/11d per hour.

ENGINEERING		COACHMAKING
Aluminium Welders	Metal Polishers	Bodymakers
Blacksmiths	Millwrights	General Machinists
Coppersmiths	*Panel Beaters	Mounters
Electricians	Pattern Makers	Painters
Examiners	Tinsmiths	Polishers
Fitters	*Tool Fitters and Smiths	Spindle Hands
*Leading Hands	Turners	Timber Markers
Machinists	Welders	Trimmers
*Magneto Fitters	Winders	

*To receive 1d per hour additional

"GROUP 2 — Asst. Craftsmen — existing rates plus bonus: 1/8d per hour.

Asst. Craftsmen	Hammermen	Tube Benders
Bench Hands	Sprayers	Tyre Press Fitters
Brush Hands	Stokers	Unit Adjusters

"GROUP 3 — General Hands — existing rates plus bonus: 1/6d per hour

Advert. Hands	Rubbers	General Hands
Bill Stickers		Mech. General Hands

"On the pay day preceding Christmas, a SPECIAL BONUS payment of 50% of a normal weekly wage will be made to all grades who are directly concerned in production at Chiswick Works.
55, Broadway,
Westminster, S.W.1.
16th December, 1926.   Signed: — Ashfield."

Prior to the initiation of this scheme the term craftsman was seldom heard in the workshop. Employees were generally proud of their skill in their particular trade, and it was an accepted principle that they be graded as such. It is interesting to note here how the word Hand was used to define the semi-skilled man, i.e. brush hand, bench hand, general hand, etc. However any prejudice as to what classification under which you were employed, was quickly swept aside when it was realised that a considerable financial gain was to be the result. The Christmas Special Bonus of 50 per cent of a normal weekly wage was the first-ever granted by the management, and needless to say it was very well received and did much to contribute to the improved relations between management and staff.

I am regraded as a craftsman

During the mid-twenties the workshop staff was regraded as we have just seen, with a revised pay structure. As a consequence of this I was now no longer classified as a fitter and turner as hitherto, but as an engineering craftsman, with an hourly rate of pay of three-pence an hour above that which I had previously received. However, I don't think anybody at that time worried one iota about what they were classified as, for it was obviously the rate of pay which was coupled with it that was really important. I continued to work on the same job as previously, on crankshaft grinding. The increased rate of pay made the job more attractive, and for a time at least it seemed in some ways to make the many and devious disadvantages less pronounced.

To keep them cool while being ground on the machines, and to wash away the dust, the crankshaft journals had a jet of water pumped continuously on to the grinding area, with ordinary washing soda dissolved previously in the water. This created a very fine spray while grinding was being carried out, and I am sure the operator's face, mouth, nose, and lungs did not benefit by it; and also what was the effect on the eyes? This caused me some concern, but it did not seem to worry the other operators, and had I complained to the supervisor I should have been told, "If you don't like the job, pack up and get out". I am pleased to say that this practice has now long since been stopped, and soluble oil is now used on these machines instead.

Once every month or two some crankshafts requiring regrinding would be received from the East Surrey Traction Co. Ltd., of Reigate. This work had to be executed after normal working hours, and hence we would have a spell of working overtime, and while we did not relish the extra hours we did all want the extra cash that was the net result. I was lucky, by the standards of this period, to be allowed to stay put on the engine section, working from time to time on all the various machines until September 1930. During that time I had witnessed many of my friends and fellow-workers receiving the sealed addressed envelope from their foreman at four o'clock precisely, informing them that the company was giving them one hour's notice to terminate their service. By five o'clock they had to collect their tools and be off the premises. One fellow among these was a regular crankshaft grinder whom I had known and worked with on and off since 1925. Over these years the volume of work would sometimes slacken off, and we would all ponder as to what was to follow, with great concern as can well be imagined, for it meant more discharges. We used to get little scraps of information through the grape vine, which were invariably correct, and perchance 20 or 50 or more bus workers would have to go.

The East Surrey Traction Co. Ltd. had been founded as an independent venture by Arthur Henry Hawkins in 1911, but became increasingly associated with LGOC, first as an agent operating on its behalf from the Reigate headquarters and from 1929 as a subsidiary. Hence many of its buses were of LGOC types and specialised maintenance work was carried out at Chiswick, including the regrinding of crankshafts in which the author was involved. This photograph of a K-type, believed to have been registered XB 8386, was taken when new in 1920 and shows the blue livery then used.

However, sometimes the work-flow would come in all of a rush, and then the entire shop would be put on overtime, working extra hours from 5 till 7 p.m., or if very busy some would have to continue work until nine o'clock. No warning was given, and you were just expected to get on with it and be thankful. If you refused to co-operate and work, it would be remembered by the power-that-be, and would reflect on you should another "clear out" be imminent. Those at our homes with meals prepared for our homecoming had to bear the brunt of this lack of thought and understanding. It must be remembered here that, unlike today, very few artisans' homes enjoyed the convenience of a telephone, indeed very few could afford the expense involved. In any case, no facilities were available for employees to phone home from within the works.

All the vehicles operated by the Company in those days were powered by petrol engines, and the mileage of these in service between overhauls would vary anywhere from about 6,000 to perhaps 40,000 miles, hence the fluctuation in the work load which had to be covered by the Engine Section. I believe at one period we were overhauling more than 40 engines per full working day, not including what we on the section called "partials". To put the uninitiated reader into the picture, and to show how things have changed, and consequently with them the volume of work involved, most modern diesel engines as now used in all London's buses can usually run 200,000 miles in service before requiring major overhaul.

In August 1930 I had special leave from the works, to enable me to fulfil my obligation to the Royal Naval Volunteer Reserve to go to sea with the Atlantic Fleet for training. This was the fourth consecutive year in which I had leave for this purpose, usually 15 to 21 days, although not always in this particular month. I would don my Naval uniform, released and relaxed from bus work, to serve for a spell on one of H.M. ships at sea or at a naval establishment ashore, and the change was very welcome. I am certain that the supervision of the Engine Section were not too pleased about the loss of my services on these occasions, but they could not do a thing about it because it was the Company's policy to encourage employees of all grades to volunteer for service with the first reserves of the Armed Forces of the Crown. This gave many bus workers a splendid, and in my estimation a worthwhile, opportunity of not only being trained and equipped to serve their country should the need arise, but also to further their experience in whatever field they chose, which in my particular case was mechanical engineering.

Soon after I had returned from this leave to my job on the Engine Section I was indeed pleasantly surprised early in September 1930 to be given the chance to make a change by transferring to the General Machine Shop. While I was away on naval service an operator from another section had been transferred to take over temporarily my position on crankshaft grinding. He liked the job, and so I was told that if I wanted to make the change I could, and he could carry on in my place. So I did vacate the job, and he took it, and, as a matter of interest, he remained on it until he retired 25 years later. During the time I had spent on the Engine Section I had gained a wealth of knowledge and experience, of the sort that one cannot gain by reading from a book in a comfortable armchair by the fireside, and I hoped it would help me in my career in bus work in the years ahead.

Refreshment while at work

The British worker is often laughed at for his liking for the traditional cup of tea, and of course the staff engaged on bus work have never been any exception to this rule. Until the introduction of the 47-hour working week in 1919 tea was brewed in the workshops individually by each man or each small group of men, with more or less official sanction, or at least a blind eye being turned. But, alas, when the shorter working week came into operation tea making and tea drinking became strictly forbidden. Both drinking and eating were now grouped together in the same category, as taking refreshment, and this was a punishable crime with no less a penalty than instant dismissal if caught so doing.

However, this curtailment of tea drinking was bitterly resented, and certainly not easy to enforce. Many employees continued surreptitiously to boil water on the blacksmiths' forges, or perhaps the welders' or the coppersmiths' blow-pipes were the medium used to supply the necessary heat. Tea was brewed, drunk, and enjoyed, but very strictly under cover and with a reliable look-out. A group of three or four employees working in close proximity would form a "tea club" or "swim", brew their tea in one large can, divide the cost, and share the risky making.

In 1927 a scheme was introduced for tea trolleys, manned by lads from the canteen staff, to be allowed on certain specified routes within the works. Here the staff were allowed to purchase 1d or 1½d worth of tea from an urn, and rolls and buns were also on sale. You had to buy your own tea and food, and the supervisors would watch the trolleys like hawks to see that too large a crowd did not gather around. Cans to contain the tea could be purchased from the canteen for 2d, and these were very similar to the beer cans in use before the First World War, except for a larger diameter at the base than at the top, and with a wire handle for carrying. Refreshment had to be partaken while continuing to work, or so the rule specified. Before the introduction of this scheme it was a real crime to be seen eating or drinking, and I have known cases of men being caught and sent home from work by the foreman. They would have to lose the rest of the day's pay, and be told to make a start again on the morrow, or perchance the day after that. Fortunately this was not always the case, and some of the supervisors would pretend not to see and look the other way.

It was not until 1930 tea kiosks were installed in various locations in the works, and an entirely different conception regarding taking refreshment while at labour was born. At these points the canteen staff prepared the tea and supplied rolls and cakes, etc., to the shop labourers, who had already obtained individual orders and payment from the various staff working in their own particular section. Each person had to supply and look after his own individual enamel mug, and paint his name or identification mark on it. Cups were not supplied with the tea by the canteen as free issue, but they could be purchased from there for the sum of fourpence each. Electric bells were installed, synchronised, and were rung to herald the start and finish of a ten-minute refreshment break. This occurred both in the forenoon and the afternoon.

Workshop salute

When a member of the workshop staff was leaving the premises prior to his impending wedding, he would be given a really warm "send-off" by his fellow workers, on the section where he was employed, giving him the "ding-dong" with the simple process of applying hammers to metal. The noise this created would be terrific, and when the bridegroom-to-be approached the adjacent section en route to the exit, the clanging would be taken up by the staff located there. This "ding-dong" would also be practised to give an employee "the bird" should he arrive extra late in the morning or perchance leave extra early. Supervisors were powerless to stop this practice for they could never see anyone banging, and the noise would always come from the fellows they could not see! Hence they always had to have a blind eye and deaf ears on such occasions as these.

Should the sound of broken crockery be heard whilst the staff were in the canteen during the lunch break, the cheer that followed, accompanied by the noise of the rattle of spoons on cups, and the thumping of hands on tables, would have been very much appreciated by many an orator, actor, or sportsman seeking applause.

Another sort of incident that created a big commotion was if anybody won any money. I have never myself been at all interested in organised gambling for money. In my early working years I had very little money to gamble with anyway, and as I grew older and wiser I realised that one's journey through life was in itself a gamble, whether you were engaged in bus work or in any other field of employment. In the late 1920's there were no football pools such as those that we know today, and certainly no betting shops. But at about this period a Sweepstake on certain classic English horse races was sponsored by the Irish Government to help the funds of the hospitals in their country. Tickets priced at ten shillings each (a large sum of money in those days) were sold surreptitiously in workshops and offices, in fact in any place where people met, all over the country, the reason for this being that they were all "supposed" to be purchased direct from the promoters in Ireland.

A hint of changes to come was given by the introduction of the LS or 'London Six' class by the LGOC. Although only twelve of these six-wheelers entered service in 1927-28, they paved the way for much larger-scale use of the subsequent LT class of six-wheeled buses. They were based on chassis built at Southall during the brief period when the Associated Daimler Co. had been set up to market vehicles produced jointly by AEC and Daimler. LS5 is seen at Marble Arch in June 1929.

5. The '30s—A Time of Change

The General Machine shop

Early in 1930 the policy of siting machines on the sections where the parts to be machined were to be assembled was largely abandoned, and a general machine shop was inaugurated on the north side of the main engineering shop. The components to be machined were transferred to this new shop on hand trucks or sack barrows from the secondhand viewers. The great advantage gained by this was that the machines were able to be used to accommodate work from various sections, and to be used for more than one type or class of work. This project marked the first step in the departure from the original mass production system which had been introduced with the opening of Chiswick Works in 1921. With the creating of this new shop all the machines installed were converted to have individual electric motors driving through endless vee belts.

The foreman in charge of the General Machine Shop had previously been in charge of all the machines in the Engine Section, and I had worked under his control for four years. Hence when in September 1930 an operator was required for a brand new Barber & Coleman hobbing machine which was being installed at that time I was given the opportunity of taking the job. It should be explained here that when the existing machines were removed from the various sections to form the new shop the operators concerned were transferred at the same time. The crankshaft grinding machines were not transferred, however, because they were of a specialised type which could be used only on this particular job, and were not of general application, so they always remained in the engine section.

I readily agreed to be transferred, although there was to be no cash benefit involved. I was heartily glad to leave the Engine Section after just over five years of service there, and I took the view that the new General Machine Shop was a good jumping-off ground for a position in the Drawing Office which had always been the first choice on which I had set my sights. In due course I started work on the hobbing machine, cutting the splines on driving shafts and on various shafts for gearboxes, also on recutting helical timing gears.

It was indeed a new experience for me to work on an entirely new machine on which no previous operator had worked, and which I could tend and operate knowing its full history right from its installation. I was extremely careful, and very cautious, for an event was still fresh in my memory which had happened on the Engine Section some 12 months previously. A new radial drill had been installed, and an operator was assigned to take over the work on it, but he had not been on the job a couple of days when he left the stop off the "self-act" device and consequently drilled a 5/8″ diameter hole right through the table of the machine. He promptly received one hour's notice to get his money and "go", and I certainly did not want an accident such as this to happen to me.

It was an interesting job, and I had the opportunity of tackling a variety of new tasks which had not previously been undertaken in our factory. One pleasing feature was that the work was reasonably clean. This shop was a happy one, and the staff in the main had been transferred with their respective machines from other sections within the works. The new machines, and there were probably eight or nine of them, were manned by operators who had been hand picked by the administration, some from within the works and some from outside. They were all a good crowd of fellows, and would go out of their way to help one another and to keep each other out of trouble. Very often there was not sufficient work here to keep one operator occupied on one particular machine full time, so some of us were given two or even three machines each to attend to at various times. I always had plenty of confidence in my own ability as a craftsman, and I took every available opportunity to gain new experience as and when the least chance presented itself. But some fellows hated this, and resented being moved from one particular job, and were quite content to carry on regardless for year after year on what seemed to me at least a most boring, repetitious, and soul-destroying task.

Soon after I was transferred to the General Machine Shop an incident happened which gave a very true picture of the workshop code of behaviour. We had a very zealous and officious attendant in the toilet close by. It was said that he had shares in the company, but whether this was true or not I don't know. However, he threatened that if he could catch any of the fellows smoking in his toilet he would report them. He tried very hard, but he had to have a witness, which somehow he could never manage. He would fetch the labour superintendent to his domain, but when he arrived the air was certainly full of smoke and yet funnily enough nobody was smoking. The men working on all the nearby sections were furious with his threats and conduct, because after all he was only a working man and not a policeman or warden, and they swore they would "have him". Sure enough, one Saturday afternoon, after he had been working overtime in tidying up, and set off home, as he passed the warden at the main gate he was stopped. The warden found he was loaded with the company's property, in the form of toilet rolls, soap, and towels, and hence that was the last we saw of him. It was obvious that the wardens had been tipped off, but by whom, nobody knew? The fellow who took over the job after this did not seem to notice the smoky atmosphere!

One day's output of Chiswick Works, July 1930. Although this picture was obviously deliberately posed, it does give an indication of the immense capacity in terms of new bodywork and overhauled vehicles. At the front are examples of the then recently introduced ST type based on AEC Regent chassis, led by ST 218. These gave better comfort for passengers, an enclosed straight staircase and spacious platform for the conductor and a livelier and more responsive six-cylinder engine for the driver, but still no windscreen, thanks to Metropolitan Police caution. A total of 829 similar buses were built, and Chiswick Works built 754 of the 49-seat bodies in the 1930-31 period. Visible behind are overhauled buses, comprising covered-top NS, open-top NS, and S-types.

Not shown are the numerous units overhauled for supply to the operating garages. Engine overhaul was probably near its peak at this time, with few of the longer-lasting six-cylinder units yet in service, so there would have been many overhauled engines in addition to those in the buses shown.

From time to time the management would have plain-clothes men wandering around the works keeping their eyes and ears open. We thought they were police detectives, or they may have been private inquiry agents. They were on the look-out for any information that might lead to the discovery of any loss of the firm's property, and also to find out who were the wicked men who took bets on horse racing from the staff. I should think that everybody knew, but when asked all seemed at that particular moment not to realize that betting took place. I once saw a fellow who we all knew was a bookie's runner look really poker-faced and blank when questioned, making out that he could not make sense of all the interrogation. But they could not catch him out, and nobody, even if they didn't like him or his betting, would give him away. After all, the rank-and-file employed in the workshops were not stupid, even if some of the management thought they were, and they certainly were not going to take on additional duties as part-time unpaid policemen spying on their own work-mates.

Until 1933 when the works was still owned and operated by the L.G.O.C. I can recall many a time seeing the Company's Chief Engineer or Operating Manager touring around the various sections within the works unescorted, and I noticed how the supervisors and lower officials would watch his every move, and run about their various sections like frightened rabbits, making it their business to bump into him at the appropriate time of their own choice. I remember on one of these occasions my Chief of the Tool Room and Machine Shop was laughing to me when he saw how some of the other Supervisors behaved, and as he stood quite still with his hands behind his back he said to me, "I am the Gaffer here, and if he wants to know anything he can come and ask me". This was typical of our own boss, for he certainly knew his job from A to Z, as we used to say in those days, and I am sure all his staff respected him. On more than one occasion when I was waiting for work to arrive, and the whole shop was short of work, he told me to "cut air" on the machine, and not to worry, for in those days when the work load of any shop began to shrink all concerned began to wonder "What next?".

Wedding bells and a move to the west

In July 1932 I decided to wed, and to dispense with all the hours of travel I had experienced for nearly eleven years since starting at Chiswick in August 1921. So I resolved to move west, to Isleworth,

In 1931, the existing fleet of 150 six-wheel open-staircase double-deckers on AEC Renown chassis delivered in 1929-30 and classified LT by the LGOC was greatly enlarged as a result of orders for 800 more vehicles with enclosed staircases of the general type illustrated opposite. This photograph taken on the assembly 'track' at Chiswick in 1932 shows the chassis of an overhauled LT about to receive its six-cylinder petrol engine and four-speed sliding-mesh gearbox.

The bodybuilding department at Chiswick spent much of 1931 producing most of the bodies fitted on the 800 LT-class buses ordered for delivery that year. There were a number of minor variations, particularly of destination indicator layout, but this view of LT 718 when new conveys the imposing appearance of these 56-seat buses.

The scene in the paint shop at Chiswick in February 1933. The body in the centre of the picture is newly-built for one of the first 100 buses of the STL-type introduced by the LGOC towards the end of the previous year. These were based on a slightly longer version of the AEC Regent chassis as previously used for the ST and were of lightweight construction so as to allow seats to be provided for 60 passengers within the stringent weight limit then in force. Note the flow method of painting being used. The other two bodies were overhauled units for LT-type buses.

and to live reasonably near to my work. During those years I had travelled many thousands of miles on the District Railway, and at no mean cost to myself, because no free travel was allowed to bus workers on the underground railways at this time as it is today. My journey from Upton Park Station extended westwards through 31 stations to Chiswick Park, and, as can well be understood, travelling this distance for twice a day six days a week became very tiring, especially after completing a hard day's work. It was impossible for me to travel to work by bus in the morning, even if I had so desired, because the journey would have taken about two hours, and the factory opened far too early, and the first bus travelling westwards from my home territory did not start early enough.

We who travelled daily to Chiswick on the District Railway from the east, all having originally worked at the Company's garages in that area, such as from "The Kings" (Seven Kings), from "The Gate" (Forest Gate), and from "The Green" (Leyton Green), were all known as "the East Enders". Coming to work in the morning we usually all travelled together in the last two carriages of the train, smokers and "nons" segregated in the two respective coaches. Some of us would have a nap, but we could rest assured that one of our fellows would give us all a shout when we were approaching Chiswick Park Station. There was a train that arrived at the station at 7.25 a.m., and if we were on this train, and were fleet of foot, we could just manage to clock on by 7.32, using the two minutes of grace that we were allocated. Hence this particular train was always known as "The Sprinters", but we usually tried to catch the one previous to this.

When any member of the staff was getting married it was the custom in the General Machine Shop and the Tool Room that a list would be taken around the two shops inviting contributions for the purchase of a wedding present. It was usually understood that our workmates would give one shilling each, and the foreman two shillings, so the sum collected generally came to about £2, plus or minus a shilling or two. The bridegroom-to-be would then be asked what present he and his intended bride would choose, and when the happy couple had decided, during a subsequent lunchtime the selected gift would be purchased and brought into the shop for all to see and approve. It was an unwritten law that on his last working day as a "free single man" the bridegroom-to-be should bring into the works a bottle of whisky, and of course a glass, for his fellow workers and his supervisors to drink his health and happiness. Naturally plenty of good humour was present, everyone being in a jovial mood. The boss of the sections, after partaking of his drink, would present the gift just prior to the recipient departing for home, which often was in advance of the normal finishing time.

Plenty of advice would be given, some good and some very questionable, and then the final send-off would be amid a terrific noise of banging and clanging of anything metallic which when struck would produce a decidedly noisy note. I should explain here that it was a real crime to bring intoxicating liquor into the works, the penalty for so doing, according to the rule book, being instant dismissal, but I must say this in fairness to the management that on such occasions as this they shut their eyes to the breaking of this rule. In my case I received as a wedding gift from my workmates an eight-day clock, which to this day is still giving me sterling service. By this time one week's paid holiday had been introduced, and the entire factory was closed during the week following August Bank Holiday Monday, which in those days was the first Monday in August and not the last as it is now. Hence this period was naturally a popular time for wedding bells to ring for the workshop staff.

I soon grew to like living in Isleworth, for it was then a quiet semi-rural area surrounded by orchards, with no noise of aircraft flying overhead, and where London Airport is today was then all open country largely devoted to market gardening. I would travel to work on a London United electric tramcar, which picked me up at the end of my road only a stone's throw from where I lived and stopped right outside the main gate of the works. The journey took about 25 minutes, and the fare paid was threepence per day as a workman's return. In fine weather I would usually cycle to work, which in those days was quite a pleasant and easy ride.

I had been married only six or seven weeks when we heard through the grapevine in the factory that some more discharges were on the way. This unfortunately proved to be correct, and before the end of September we said good-bye to some more of our workmates, although we in the General Machine Shop lost only one fellow. I knew one of our "East Enders" who was sacked at this time, and although he was a skilled man and an ex-apprentice of the company he was now glad to suffer the indignity of accepting a job sweeping railway station platforms for the princely sum of two pounds a week. What a contrast to the situation when I retired, when really appealing advertisements for staff in all grades were being displayed on the interiors and exteriors of buses and properties belonging to London Transport and in the national and local Press, with apparently very little response from qualified and acceptable applicants.

Towards the end of 1932 the air in the whole factory was thick with rumour as to what was going to happen when the company was to be transferred to the new public authority, which we learned later was to be known as the L.P.T.B. One would have thought the management would have kept their work force informed of the likely events which they must have known would have to follow, but no, as usual, all the employees knew as to what was going on were the little scraps of information which appeared from time to time in the Press. All we could do was wait, tense and apprehensive, as to what changes would be made and how they would affect us as individuals, and whether working for a public authority would prove an advantage to us all as compared with working for many years for a private enterprise company as so many of us had done.

Although by far the largest operator, the LGOC was by no means alone in operating buses on London's streets. Here, GK 891, a Leyland Titan TD1 of the Premier fleet is seen in Trafalgar Square on 12th March 1931. It was later to become London Transport's TD 67.

London Passenger Transport Board, 1933

Section One of the London Passenger Transport Act, of 1933, provided for the establishment of a public authority to be known as the London Passenger Transport Board, consisting of a Chairman and six other Members, to be appointed from time to time by a body to be convened for this purpose by the Minister of Transport, and defined and referred to in this Section of the Act as the Appointing Trustees. In accordance with the principles and procedure laid down in the Act, the appointing trustees duly appointed the Chairman and the first Members of the Board, on 18th May 1933. The Rt. Hon. Lord Ashfield of Southwell, was appointed as the Board's first Chairman. He had already been in control of most of London's buses for the previous 20 years and most of London's railways for the previous 30 years, being the Chairman and Managing Director of the Underground Electric Railway Co. Ltd., its four subsidiary companies which operated the Tube and the District railways, the three large tramway companies operating in the London area, the London General Omnibus Co. Ltd., and the Associated Equipment Co. Ltd. (which had been set up in 1912 to take over the manufacture of buses from the L.G.O.C. though it had meanwhile built up a flourishing business making buses and lorries for sale to other companies).

The duties to be undertaken, and the objects to be achieved, by the new board were defined in Section Three of the Act. By Sub-Section One of that Section the Board was charged with the general duty of providing an adequate and properly co-ordinated system of passenger transport for the London Passenger Transport Area, as defined, and for that purpose they were required, while avoiding the provision of wasteful competitive services, to take from time to time such steps as they considered necessary to extend and improve the facilities for passenger transport in that area, in such a manner as to provide most efficiently and conveniently for the needs thereof.

The London General Omnibus Company Limited, which was a part of the Underground Group of companies, was duly transferred to the London Passenger Transport Board on 1st July, 1933, as also were the Tube and the District railways, and the three electric tramway companies in the Underground group. The A.E.C. was excluded, however, and thereafter remained separate, but the Metropolitan Railway Company and 12 municipally-owned tramway undertakings were taken over on the same date, and about 140 small bus and coach undertakings followed gradually during the next 2½ years either wholly or partly. Consequently all the employees of the L.G.O.C. at Chiswick came under the control of the new Board on and from 1st July, 1933.

At first little change was apparent, and the routine of the works continued as before. But gradually the engineering employees of other undertakings absorbed by the board were transferred to Chiswick, and placed in positions comparable to those they had previously held elsewhere. While appreciating that this procedure was fair to them, it did in no small way stop the anticipated promotion of many of us who had been employed in the factory for many years.

Another Premier Titan, GC 1214, overtakes a General NS on Ludgate Hill, almost under the shadow of St. Paul's Cathedral, on the same date in 1931. Some independent operators were able to take advantage of the markedly superior performance of such buses as compared to the older generations of LGOC types to provide faster services. The body designs were influenced by Metropolitan Police requirements in both cases—windscreens had been virtually universal on provincial buses for about a decade. When the new Board took over all London bus operations, the variety of makes and types created problems from a maintenance and spare parts viewpoint.

All the buses transferred to the newly-formed London Passenger Transport Board had to carry its name as legal owner. The ex-LGOC buses were not changed in livery, and even the 'General' fleetname continued for a time, but it was quite a task to change the statutory lettering so that they would all be carrying the correct name on the first day of operation. Here a transfer is being applied to an overhauled bus.

We were now in the unfortunate position of having many new-comers transferred into key posts, and being tutored by experienced men who themselves could have reasonably expected to be promoted at some future date. The formation of the Board did little to enhance or improve in any way the position or conditions of the former employees of the L.G.O.C. Many of us had been misguided enough to believe that following the setting-up of this new autho-rity our conditions of employment were to be greatly improved. This was not so, and many employees suffered disappointment that this co-ordination of bus work which had previously been given so much publicity, and which had banished competition between operators, had achieved little to improve the conditions of the employees. In many cases the conditions of transferees were worsened, as many now had to travel long distances to reach Chiswick Works, notably from the Lewisham and Catford area, and from the Reigate and Redhill area, a repetition all over again of my own circumstances from Forest Gate 12 years previously.

The Board took over many types and makes of vehicles from the constituent undertakings, and it was their policy from the outset to maintain standardisation in road vehicle design, to secure economy in maintenance and overhaul. In this respect they continued the L.G.O.C. train of evolution unchanged, and ignored all outside influences from acquired undertakings though of course these were much smaller. The Board's initial fleet of some 5,000 buses and 400 coaches was composed almost entirely of petrol-engined vehicles, the vast majority being fitted with crash-type gear-boxes. Some 1,837 of the 3,940 L.G.O.C. double-deckers, or a trifle more than half, were of the NS type, mostly with outside staircase, which had first been operated by the L.G.O.C. in 1923, and the other half were mostly of the ST and LT types introduced in 1929/30 culminating in the STL type, which had been introduced by the L.G.O.C., and of which the first 150, in three separate batches of 50 each, were ordered by them, and 75 were in service by 1st July, 1933. The Board continued with this new STL design at the rate of about 500 per annum for five years, though the later vehicles incorporated further important design changes. The fleet acquired by the Board from the General, excluding the General's important country services and Green Line subsidiary companies, whose vehicles were maintained at Reigate, but including the Overground subsidiary, whose 52 vehicles were already dealt with at Chiswick, was as follows:-

Double-deck:- 1,837 A.E.C. "NS", 12 A.E.C. six-wheeled "LS", four L.G.O.C. six-wheeled "CC", 1,222 A.E.C. Renown six-wheeled "LT", 814 A.E.C. Regent "ST", 75 A.E.C. Regent "STL", 25 Dennis Lance "DL", and 3 Daimler "DST", total 3,992. Single-deck:- 47 A.E.C. "S"-type, 3 L.G.O.C. "CB", 45 A.E.C. Regal "T"-type, 199 A.E.C. Renown "LTL", 4 Dennis "DS", and 1 A.E.C. side-engined "Q"-type, total 299 full-size types; plus also 40 Dennis Dart, 7 Guy, and 3 Bean small 14-20 seaters.

On the technical side, the Board was, from the start, aware of the advantages of the oil (diesel) engine as compared with the petrol engine, particularly from the point of view of economy in operation and maintenance, so they accordingly adopted the oil engine as the standard power unit for all new purchases from 1934 onwards. The L.G.O.C. had been one of the pioneer concerns in this field, and had three ST type buses briefly fitted with oil engines from November 1930, and more importantly, 28 LTs from 1931 and another 75 from 1932.

Before the formation of the Board the small independent opera-tors were often wrongly referred to by the L.G.O.C. bus workers, and indeed by the general public and especially the newspapers, as "pirates". This was quite wrong, because they had to pay their taxa-tion, licences, and insurances, etc., just as much as the General, but I think it was understandable, because the small firms tended to pick the most lucrative routes and the best times for running their buses.

The largest operator other than LGOC taken over by the Board was Thomas Tilling Ltd., with 369 buses. From the early days, until 1930, the fleet had been standardised on Tilling-Stevens petrol-electric vehicles, such as this ten year old TS7 model. It was photographed about the time LPTB was formed in 1933, and by coincidence the General bus following also had a transmission system designed to give smoother operation and easier driving. LT 1335, an AEC Renown six-wheeler built in 1932 was one of the early batches of vehicles with a fluid flywheel and preselective epicyclic gearbox, the system that was soon to be standard on London buses and formed the basis for later semi- or fully-automatic transmission. The Tilling-Stevens is on service 1, which is still operated by Catford garage, whose Tilling origins are still indicated by the TL garage code seen on the waistline just behind the driver. Such vehicles were both non-standard and rendered completely obsolete by the rapid design changes of the time and called for rapid replacement by the new Board.

In the main they gave very good service, and certainly provided some healthy competition for the L.G.O.C. where they ran on the same routes, although many of them also inaugurated some quite new routes.

The largest bus operator apart from the General was Thomas Tilling, Limited whose 369 buses were taken over on 1 October, 1933, and their overhauling immediately transferred to Chiswick. On the same day we also received a further 33 of NS type from the "British" fleet. The former East Surrey (by then London General Country Services) and Green Line fleets remained self-contained at Reigate for nearly two years, absorbing in the meantime about 85 other operators, some of them quite large, but eventually their vehicles were gradually transferred to Chiswick, the standard type coming first and the miscellaneous oddments later. Meanwhile some 52 other small or moderately small fleets, all of which had operated in London itself as distinct from the country area, had been taken over directly by Chiswick.

We in the factory at Chiswick soon got used to seeing all sorts of unfamiliar vehicles painted in strange exterior liveries entering the works for overhaul or sometimes to be fitted with equipment to bring them up to the Chiswick standards. Likewise in the shops we would see units and components from hitherto unseen makes of chassis being overhauled. The work-force in the establishment grew with the arrival of men from acquired undertakings, but likewise so did the volume of work and in some cases the type of work involved. In the general machine shop we had three fellows who came from the East Surrey and one from Thomas Tilling. They were quite good chaps, and very soon settled down with the rest of us in what to them was a huge factory very different from those from which they had been transferred. The fellows from Tillings were known to us as the "Bull Yard" men, because their overhaul works had been at the Bull Yard, Peckham, and those from the East Surrey were known as the "swede gnawers" because most of them lived in the country villages surrounding Reigate.

While I freely agree that these two groups of men, when they were transferred to Chiswick, had to travel in very many cases very long distances to work every day, nevertheless the conditions that they enjoyed and the pay they received were infinitely better than

London Transport's domain included the country area surrounding the metropolis and here, too, independent companies and their vehicles were transferred, although this took place more gradually in 1933-35. In some cases, their choice of vehicles conformed to the standard inherited from LGOC so far as the make of chassis was concerned. Amersham & District had placed a number of AEC Regals on the road, for example, and in due course these were added to the ex-LGOC T class. The bodywork was quite different, and the choice of Strachans as bodybuilder and the livery and style of fleetname closely followed the practice of Aldershot & District. KX 7634 eventually became LPTB T361, and subsequently received an oil engine and new LPTB-design Weymann body.

they had previously had. For my part I regarded many of them as a very docile and servile crew, with far too many "yes men" in their midst. Most of them were "one job" men, who thought that the way they were accustomed to undertake a particular job was the proper way. Many had not advanced with the modern methods and the use of the numerous new machines and equipment which we had at Chiswick. I agree some were very clever, experienced, and ingenious craftsmen who had learnt how to accomplish jobs skilfully with very little up-to-date machinery or equipment. Some of them would say, "We used to do so-and-so at our old place", but they were promptly told, "Forget it, you are at Chiswick now", and they were soon brought into line with the Chiswick code of conduct; indeed they had to or they would soon have been in trouble.

Some human aspects of factory life

A wide variety of miscellaneous goods were sold privately in the works by the staff from time to time, and anything from razor blades to cameras or shirts and socks could be purchased very much cheaper than the standard retail price. Of course all this went on strictly under cover, and no questions were asked. How the goods were obtained and made available nobody knew, or for that matter cared, the important feature was that they were cheap. I leave the reader to guess how they were obtained; perhaps they "fell off a lorry?

Raffles also were quite common at this time, and such items as watches, clocks, jewellery, etc., which the average bus worker could not normally afford for cash down at a moment's notice, were offered to those fellows who were interested in a chance to win the item by purchasing a ticket for 6d or 1/-. The draw for the raffle would take place in the workshop during the dinner hour, for all interested to see. An official Christmas Draw was run every year by the Chiswick General Sports Association, and I believe the proceeds went towards the cost of the annual Children's Party, which was held each year in the works canteen and still is today.

One sombre event that used to occur in the canteen from time to time, in this period when money was short, was a sale of the tools which had belonged to a deceased workmate, in a small effort to raise a little money for his dependants. These sales would be well supported by the rank-and-file. It should be borne in mind that in this era very few tools were supplied by the management, and most craftsmen such as toolmakers, pattern makers, and coachbuilders had to spend many pounds in equipping themselves with a suitable and comprehensive kit of tools to be able to follow their trade. The tool kits were of course added to as and when the need arose and the necessary cash was available, and many of the coachmakers in the body shop had large cabinets full of their own personal tools.

The widow's pension from the State at this period was only 10 shillings per week, plus a few shillings for each dependent child. So a penny collection would be collected from everybody within the entire works, though you could give more if you so desired. This had the management's consent, and after a small floral tribute had been purchased the rest of the collection would be given to the widow and dependants of the departed work-mate. I think everybody gave this small amount, and considering that at this time the works employed well over 3000 men, quite a good sum of money by the standards of those days was collected. The company, and after 1933 the Board, gave the dependants precisely nothing to my knowledge, but to be fair to them I understand that it was their policy to offer employment to the widow in whatever sphere she was qualified.

Many of us had thought that following the formation of the Board the workshop staff would now be kept informed of the changes that would be taking place from time to time over the years to follow, but unfortunately this turned out not to be so. Strange fellows would arrive in the various shops, and wander about, and nobody knew who they were or what position they held in the organisation, until somebody would ask their supervisor and be told he was the new head of so-and-so section, or perhaps a new progress clerk, although sometimes even the supervisor did not know. All strangers and newcomers were treated with the utmost suspicion, especially if they were well-groomed and did not arrive in the factory until after 9 a.m. office hours. No questions were answered, and nobody knew anything, for nobody was going to risk being caught, so all queries and questions were referred to the foreman. Why any changes in organisation were not announced on the notice boards by the time clocks nobody knew.

In 1936 we had a change of the section engineer who was in charge of the tool room and the general machine shop. Now this gentleman had the right idea, and one which at least appealed to me, and seemed the correct action to take when first appointed to take charge of a group of men. He came round to every man employed in the two sections, and introduced himself as the new "boss". He asked each fellow his name, and the job he was doing, and then proceeded to ask for their co-operation and support in making the shops an efficient and happy part of the establishment. I am sure all the fellows appreciated this action, and I am confident it paid good dividends to the management. Why couldn't all the other persons holding responsible positions have taken some action such as this? In those days, alas, I think it was thought that dignity was at stake in conversing to ordinary bus workers unless it was absolutely necessary.

Nowadays carefully chosen members of the administrative and control staffs are sent to a management course. But by what I have heard from friends and colleagues who have attended them, these courses seem to be largely a social get-together, a waste of time and money, and do very little to improve staff relations or management problems that could not be solved by sound judgement and common sense. What a contrast to the early days of Chiswick in the 1920's, when the bus workers in the factory were in my estimation really bullied, with plenty of "Here you, do this", or "that", with simply no consideration at all for the person to whom it was addressed. It was really taking full advantage of the employment position at that time, for this was the period when the employees had to keep their mouths tightly shut and bite their tongues when they really felt inclined to tell the management just what they really thought of the staff relations.

At that same time the attitude of the management towards their customers, the passengers, was entirely different. A notice which was displayed in the interior of the B-type buses 50 to 60 years ago illustrates this point very clearly, for it read:- "Passengers are respectfully requested to abstain from the objectionable habit of spitting", whereas now you have curt notices such as merely "no smoking", "Used tickets", or "Leave by centre doors".

There was one little human aspect of factory life that rather amused me, when one of our chaps used a bit of private enterprise. As I mentioned earlier, the staff eligible for free travel passes were required to have these renewed at the beginning of every New Year, as all were issued annually and valid only until 31st December. Each employee who had a pass had to hand in to the Works Labour Bureau about a fortnight prior to the New Year a personal photograph of head and shoulders, measuring 1in x 1.1/8in, with his name and clock number written on the back, which the Office staff would then affix to their particular Pass for next year.

I remember one year one of our fellows used his initiative, with the idea of making for himself a "bob or two", by taking photographs of all the men he could muster during various dinner-hours. This idea proved very popular. For one reason it was cheaper than photos taken commercially, which was very important to us, and for another thing it was much more convenient and time-saving. I laugh even now when I think of it. To save time this fellow would

London Transport has always tended to be associated with large city buses, but one of the first LPTB projects which the author's colleagues in the Chiswick drawing office were involved was the development of a new 20-seat bus on Leyland Cub chassis for country and suburban routes. Cub models were produced at Kingston-upon-Thames, only a few miles from Chiswick. The prototype C1 is seen here on the 237 service at Sunbury Cross in November 1934.

line up all his customers into a single line backing onto a wall at the rear of the works, making it look like a setting for an identification parade, or even an execution, and he would then proceed along the line and "snap" each customer in turn. I must explain that in those days the company had very strict rules regarding the taking of photographs in any part of the works. These portraits could hardly be regarded as a serious infringement, but nobody was prepared to take a chance, and so a look-out was posted to give warning of the approach of anybody considered important in the works management.

Visits to exhibitions and shows

With the opening and development of Chiswick some of the more serious-minded apprentices who had been transferred there conceived the idea of visiting some of the various exhibitions that were held from time to time, usually at the Olympia in Kensington. These covered various subjects, of which quite a number were in some respect allied to bus work. Some of those which a fellow apprentice friend and myself always made a special point of visiting each year were the Motor Show, the Commercial Motor Show, the Cycle and Motor Cycle Show, and the Machine and Tool Exhibition. Quite a range of useful information could be gained from visiting such exhibitions as these, which could well prove very useful and beneficial to students and would-be engineers.

But alas, for us apprentices to go during the daytime was quite out of the question. Firstly we would not have been allowed time off from work, but even if we could have been we would not have received any pay for the time thereby lost, and which none of us could afford to lose. I have no clear idea now what the price of admission during the daytime was then, but I suppose certainly somewhere between five shillings and seven and sixpence. This gave young men such as ourselves three hours in which to obtain all the literature and data we required, and to see as much as we possibly could of the exhibits. We would leave just on nine o'clock, closing time, which then for me meant a journey of an hour and a half to reach home, and then of course, to have to be up again early on the following morning so as to be at our local railway station by six a.m., in order to be sure to reach Chiswick by starting time.

In all my many years of service in the various workshops on only one occasion was I offered a free ticket for an exhibition or the like. This was for a film show and lecture on the various processes in the

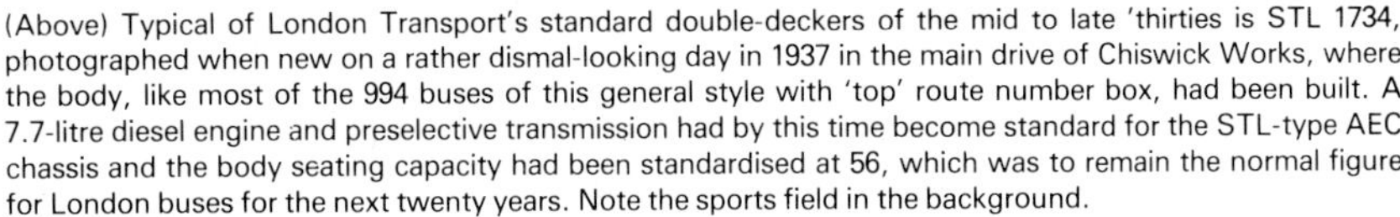

(Above) Typical of London Transport's standard double-deckers of the mid to late 'thirties is STL 1734, photographed when new on a rather dismal-looking day in 1937 in the main drive of Chiswick Works, where the body, like most of the 994 buses of this general style with 'top' route number box, had been built. A 7.7-litre diesel engine and preselective transmission had by this time become standard for the STL-type AEC chassis and the body seating capacity had been standardised at 56, which was to remain the normal figure for London buses for the next twenty years. Note the sports field in the background.

(Right) But not all the new buses were built in London. Leyland secured an order for 100 Titans with Leyland-built bodywork, also delivered in 1937, and classified STD by the LPTB. Though based on the TD4 model of the time they incorporated many special features. One is seen being inspected prior to delivery, with buses for Ledgard of Leeds and Central SMT visible in the background.

manufacture of grinding wheels for engineering machinery, and data relating to wheel revolutions, work speeds, and traverse. Needless to say I duly attended, for I could not let it be quoted that I had promptly refused the only ticket I was ever offered. At this period I was operating a Landis crankshaft grinder located in the Engine Section, and the subject of grinding wheels was interesting, absorbing, and beneficial to my career.

After the Second World War was over I had by then been appointed to the technical staff, and was working in the Drawing Office as a draughtsman. By now various exhibitions were being held again, as before the war. The Commercial Motor Show was now held biennially, and all the technical staff were required to visit it. We had to attend on pre-allocated days, receiving full salary plus expenses, and to make a special study of one particular subject or subjects that were assigned to each of us. On the following day we had to prepare a written report containing any suggestions which we thought could be helpful to our Organisation. I used to relish writing these reports, as they gave me an excellent opportunity of expressing my own ideas and opinions, which were not necessarily the popular ones.

A call for jury service

Early in December 1936, when I was working in the General Machine Shop, on arriving home one evening, I found a long blue-coloured official-looking envelope which had arrived by post that morning. On opening it I discovered that it contained a Summons for me to attend the No. 1 Court at the Law Courts in the Strand, for jury service on 1st January 1937.

I showed the Summons first to my foreman, and he said "Take it over the road and let them sort it out". ("Over the road" was the term which everybody in the works used to describe the office block.) So I went over to the Works Labour Bureau, and saw the Office Superintendent. I showed him the Summons, and he stated "it would be allright for me to go", trying to imply that he was approving of my attendance when in fact I was well aware that he dared not stop me. He then informed me that each day I attended the Court I was to ask the Clerk of the Court to sign on the back of the Summons to verify that I had attended.

I duly went to the Court on the morning, as requested, well on time. However, my name was not called, and together with the other would-be jurors whose names had also not been called we were all instructed to attend the Court again at the same time on the morrow. On hearing this I made my way to the Clerk of the Court and very politely asked him if he would sign my Summons for my employer, so as to certify that I had attended there that day. He reacted by nearly blowing my head off; was he annoyed! He said, "Who are your employers who would dare to dictate to a Clerk of the King's Court?". I replied, "The London Passenger Transport Board". He then said to me, "What will happen if I refuse to sign your paper for you?", and I replied, "I shall very probably not get paid for my loss of earnings whilst coming here". He then said quietly to me in a kindly voice, having now overcome his anger, "I will sign your paper for you when I release you from Service, but in the meantime you will attend this Court every day until I tell you to stop". His final remark to me was, "I'll teach your bus people to try to dictate to me".

I subsequently returned to the Works, and went to see the Labour Superintendent, and told him that the Clerk had refused to sign my form. But I conveniently forgot to mention that he had told me on the quiet that he would sign at the completion of Service. Was he annoyed! He told me that this had never happened before, and it was obvious to me that none of our fellows had ever previously been assigned for jury service at this particular Court. In any case only house owners at that time were liable for this task, and although I did then own my own house most of my colleauges were only renting theirs.

So I continued to attend each day, repeatedly not being actually called for service, until Thursday 14th January, when at last I was indeed called to serve on a jury. The case proceeded nearly all day, until late afternoon. It was an action brought by a professional trumpeter against his former employer for assault and damages for the loss of two front teeth which had been knocked out by a blow at his place of employment following an argument. He claimed that without his teeth he would never be as good a trumpet player. Many years previously I had been a solo bugler in the Boys Scouts, and I knew how very important those two front teeth are for playing any sort of wind instrument, also that a denture is not suitable, and hence I said the damages ought to be at least £1,000. After further deliberation, we, the jury, fixed this amount at £1,531, which was a lot of money in those days, so the trumpeter had won his case handsomely.

When this was announced the Judge closed the Court for the day, and the Clerk told all of us who had served on the jury that we were now discharged, and exempt from further jury service for some years. On hearing this I went to see the Clerk of the Court and asked him if he would now like to sign my form for me. Of course, he remembered me, and said, "Ha, you are the bus worker who wanted a paper signed. Have you had a good time here, and a rest from work?" I replied "Yes", and he then asked, "Have you had enough time at the Court?". I told him I would like to get back to my work now and he replied that it would be a pleasure to sign my form, whereupon I thanked him very much and retired from the Court. On returning to Chiswick I handed the form in to the clerk in the Labour Office, and I am pleased to say that London Transport eventually paid me in full for my loss of wages through attendance at the Court, which was very generous of them as they were under no obligation to do so.

My move to the Tool Room

In April 1938 an almost unheard-of event occurred! One of our fellows in the Tool Room gave notice of his intention of leaving in a week's time. I soon heard of this, and I applied for the job by a letter addressed to the Labour Superintendent. Mind you, I had been asking for a chance to make this change for almost as long as I had been working in the General Machine Shop. So I was thrilled when I received a note from the Labour Bureau, signed by the Production Engineer, informing me that as from 24th April, 1938, I was to be regraded as a toolmaker, at the rate of 2/- an hour instead of 1/11d. I might add that I had previously received a "whisper" from the

The Tool Room in Chiswick Works. Though this photograph was taken in 1950, there had been little change from 1938-39, when the author worked on the bench on the right. Precision engineering methods were becoming increasingly important as vehicle designs developed.

foreman that I was to be promoted. It seems ridiculous now that all I was to receive for all the extra skill and responsibility that I was to undertake was one additional penny an hour over the craftsman's rate.

However, I was happy to take the position, and I very soon settled down to the job. I had known all the fellows there for years, working as we had been on adjacent sections. I liked the variety of work and the conditions. We were left alone on whatever job we were doing, to manage for ourselves without pressure from other people, and a whole range of jigs, tools, cutters and gauges were all made here for the works and the garages. The old too-familiar cry of "more output", which was so often heard on the sections, was certainly not heard here. We all had a job to do, and the standard of work was extremely high, as is essential in all tool rooms. All the fellows here worked in harmony, and very conscientiously, for the Tool Room staff were, and indeed still are, key men in the whole establishment.

I have often heard fellows in offices talk and talk and talk about projects and jobs to be undertaken, and what should be done to achieve the desired result. If only talking could do the job it would soon be accomplished, but you cannot talk yourself or indeed anybody else into producing an accurate jig, gauge, or press-tool, etc. To be able to talk well is not essential, but you must have the skill, ability, and experience. I was very happy working in the Tool Room, and more contented there than I had ever been since I had completed my apprenticeship.

Let me tell you of an incident which gives some indication of the good atmosphere which prevailed between our Chief and the rank-and-file in the Tool Room. We all made our own tea in cans of water boiled with a gas blow-pipe which was installed on a bench, with a part fire-bricked surround, primarily for the staff to harden and temper small tools and punches. The Chief and the foreman certainly knew this went on, but provided the fellows kept it well under cover nothing was said. But one day one fellow, I guesss he was getting careless and not keeping his eyes open wide enough while boiling the tea water, suddenly found the boss almost on top of him. He had to move a bit sharply, and consequently leave the

As well as supplying the remainder of the works with jigs, tools, cutters and gauges, many items of these types were made by the Tool Room at Chiswick for London Transport garages. This is a contemporary scene at the Hertford garage, housing part of the green 'country bus' fleet. All but one of the vehicles receiving attention were Green Line coaches. Nearest to the camera is Q202, one of the coaches based on AEC Q side-engined chassis. Next is one of the T-class AEC Regals from independent operators fleets that had been fitted with new bodywork to LPTB design. Then come three T-class coaches, two examples of the early petrol-engined fleet dating from 1930-31 flanking one of the 1936 vehicles with diesel engines generally identified by their classification 9T9.

can with the blow-pipe still burning. The boss simply folded his arms and stood there. The water boiled right away, then the can collapsed, and then in turn it burnt to a charred heap of metal, whereupon he then turned off the gas and walked away. But he was a good fellow, and I am sure he did this only so as to teach the men to be more on the alert and not to take liberties.

We had one fellow working with us who was an extremely nice and amicable person, but whose one very apparent weakness was that he simply could not keep any information to himself. Hence if anyone wanted something or other spread around the shop quickly, such as a rumour, a bit of scandal, or perchance a hint of some overtime in the near future, they would promptly tell this chap, with of course a word in his ear for him to keep it strictly "under his hat". Consequently in a very little time everybody in the shop knew exactly what the person concerned wanted them to know.

I remained working in the Tool Room amicably on bench work until March 1939, when I was asked by our chief if I would care to work in the Metal Shop, still remaining under his control as one of the tool room staff, but doing the tool setting and general tool work required. there. This vacancy had arisen because work in the Metal Shop was increasing, due to the extra work involved in producing all the numerous and various metal components required for building the new RT2-type double-deck bus bodies, which contained far more metal and less wood than any previous type. I had enjoyed the year I had spent working in the Tool Room, and I had certainly learnt a great deal there, and had gained a wealth of experience that I could not have received elsewhere in the factory.

But I was also for ever game to grasp every opportunity to seek new experiences in whatever field of bus work that became available to me. So I accepted this new job, and forthwith commenced work in the Metal Shop, where, as in the Tool Room, we were treated more as individuals and not just as persons bearing a time clock card number. Little did I then know, that with the exception of three months in 1941 which I spent in the Tool Room of the London Aircraft Production Group at Chiswick, I was to remain working in this Metal Shop throughout World War Two.

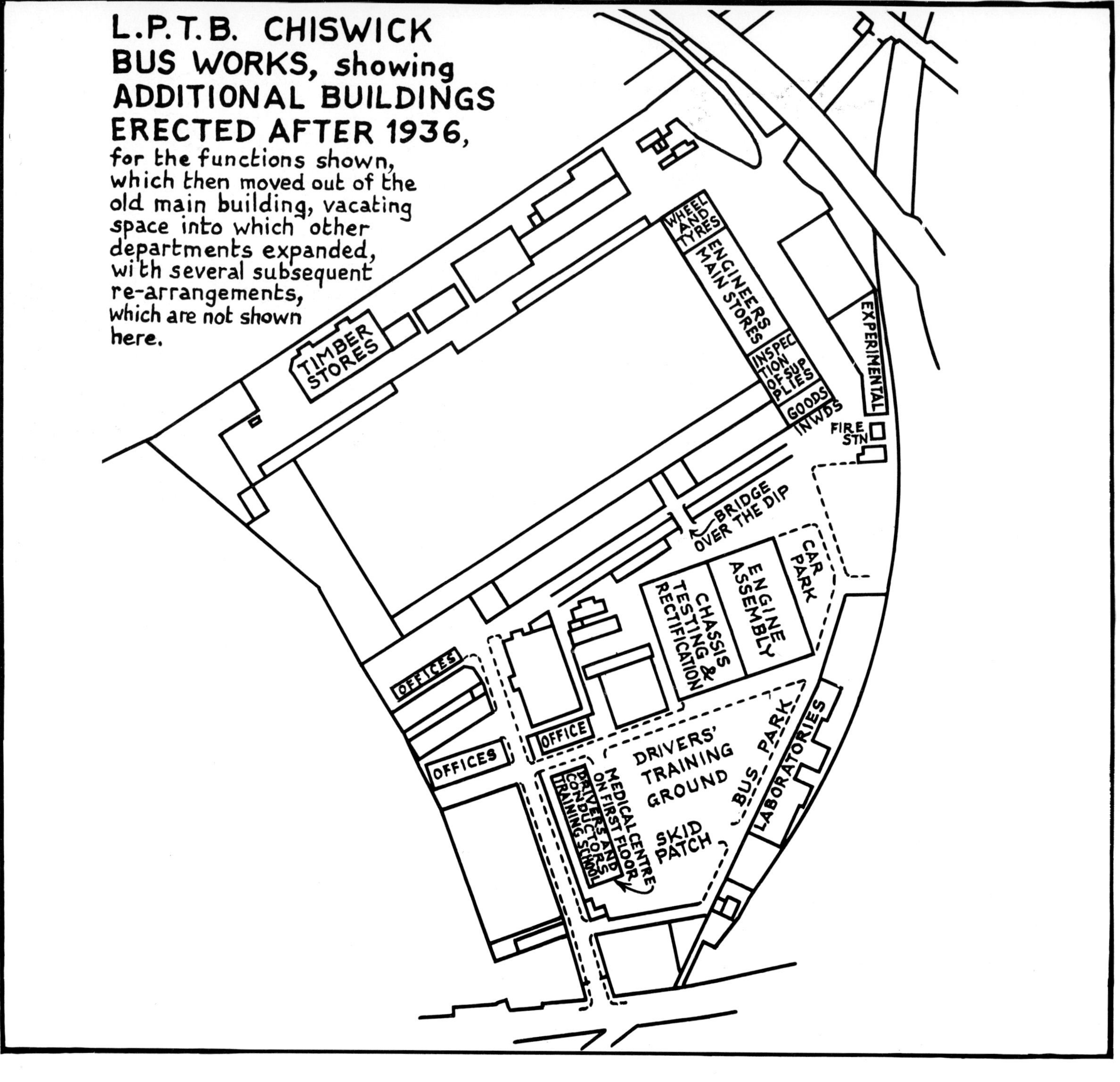

L.P.T.B. CHISWICK BUS WORKS, showing ADDITIONAL BUILDINGS ERECTED AFTER 1936, for the functions shown, which then moved out of the old main building, vacating space into which other departments expanded, with several subsequent re-arrangements, which are not shown here.
TIMBER STORES
WHEEL AND TYRES
ENGINEERS MAIN STORES
INSPECTION OF SUPPLIES
GOODS INWDS
EXPERIMENTAL
FIRE STN
BRIDGE OVER THE DIP
CAR PARK
ENGINE ASSEMBLY
CHASSIS TESTING & RECTIFICATION
OFFICES
OFFICES
OFFICE
MEDICAL CENTRE ON FIRST FLOOR, (DRIVERS AND CONDUCTORS TRAINING SCHOOL)
DRIVERS' TRAINING GROUND
SKID PATCH
BUS PARK
LABORATORIES

The RT-type bus

Early in 1936 it was decided by the Board that the technical staff of the chief mechanical engineer (Road Services), in conjunction with that at A.E.C., should be given the task of designing a new vehicle to supersede the ST, LT, and STL types which at that time comprised almost the whole of the double-deck fleet. In due course the 1RT1 design of double-deck bus was evolved, and a prototype chassis was built at the A.E.C. works at Southall by November 1938. It was equipped with air-operated pre-selector gearbox and fluid-flywheel transmission, air-pressure brakes and larger engine than any previous London bus. Many other new features were embodied in this chassis, which was well in advance of all other bus designs in operation at that time. The chassis ran in public service for three months, hidden underneath an old secondhand body, and meanwhile the prototype new body which was being built at Chiswick was in due course finished. This also embodied many new features, and was a great advance on all previous bodywork designs. The new body was mounted on to the new chassis in March 1939, the vehicle thus receiving the fleet number RT1, and after a further three months of technical tests and trials the prototype vehicle entered public service in July.

It was immediately successful, and it had already been decided to build a further 150 such vehicles in the two factories in Southall and Chiswick. The production batch incorporated many minor improvements to both chassis and body, and was designated the 2RT2 type, i.e. 2RT chassis and RT2 body. The chassis was fully extended behind the rear axle, as was still the standard practice until that time, but unlike RT1, in order to support the rear platform of the body. I personally was closely involved in building these bodies, but I did not have much contact with the chassis, which came in from A.E.C. The body was of composite structure, and embodied flitched-timber pillars with steel box-section waist rails. The front bulkhead was the first of all-metal welded-construction section ever fitted to a Chiswick-built double-deck vehicle. This bus was also the first vehicle to be fitted with window pans incorporating Simplastic glazing. Another innovation was the use of 3ft - 11in pillar centres, known as wide bay construction, which was an increase of more than 10in on all previous London bus designs, and resulted in four large windows on each side of the main saloon instead of five smaller ones as hitherto.

The layout of the body shop and the metal shop at Chiswick were extensively altered to accommodate the work involved in building these RT2 bodies. All the sheet-metal panels and brackets and presswork and other steel or aluminium details were produced in the main Metal Shop, while tubular seat frames, stanchions, grab

Some 40 years of continuous service to the public was given by the RT type of bus. The prototype vehicle, RT1, shown here in original condition, set new standards of design and appearance, and was completed in March 1939 — the last examples of the type were taken out of service on 7th April 1979. The original body of RT1 has been preserved, mounted on a post-war chassis which has been given the registration number EYK 396 originally carried by RT1, the chassis of which was scrapped in the early post-war period.

handles, etc, were produced in its subsidiary that was known as the Chair Shop. While this programme was being pursued the ordinary "bread and butter parts", as we called them, for maintenance of older service vehicles, were of course still being produced. The RT2 work started early in 1938, and very soon settled down to an even and regular output, and the 150 bodies were completed during November 1939 to June 1940.

This project was something new, which the metal shop had not previously been called upon to undertake in such magnitude, because all the many thousands of previously Chiswick-built bodies had contained far more timber and far less steel or aluminium, indeed hardly any metal at all other than panels and grab-handles, though metal framed units had been introduced in 1936. The enthusiasm with which the project was tackled was very pronounced.

The Metal Shop, in June 1939. The author was at that time working as the tool setter for the row of presses on the left of the picture and was to remain there, often on night shift, throughout the 1939-45 War. The machines to the right are guillotines.

This could be understood when it was common knowledge through the grape vine that the order in hand for 150 vehicles was only an initial one, and many similar and larger orders were expected to follow. Unfortunately these bodies were almost the last ever to be built at Chiswick, due to the outbreak of war in September 1939.

It was really very interesting to note the team spirit and keen interest manifested by all the people engaged on this RT2 programme. I was one of the five toolmakers who were made responsible for the condition, setting, safety, and operation of all the machine tools, power presses, multi drills and punches, power cutting machines, and guillotines in the entire shop. We also had to give the necessary instructions and guidance to the operators to enable them to produce the desired products. Each toolmaker had his own section of machines, and it was his task to keep them running with as little delay as possible. This was the first time as yet that I had been engaged on a job that was predominantly body-work as opposed to chassis or mechanical work. I found it a very interesting job, and it brought me into close contact with the people in the Body Shop and the Saw Mill. The personnel of the Metal Shop were the finest group of fellows I have ever been called upon to work with. Occasionally some unorthodox or unofficial methods or procedures had to be employed to produce the desired production result, and in all such instances the co-operation of everybody could be relied upon and was indeed freely given.

Night Shift

While the production of the RT2 double-deck body was proceeding through the factory, on some occasions the assembly section was "near waiting" for certain specified metal parts. When one reflects on the multiplicity of parts made of metal which were required to produce the assembled body it certainly reveals the good techniques, planning, and co-ordination which existed between the production groups as the various stages of production took shape. The machine section of the Metal Shop was working to full capacity. All its machines were fully manned and operating, but in some instances were still unable to produce all the output required. So it was abundantly clear that there was only one remedy — to start a night shift!

Hence, as a tool-maker-tool-setter, located in this shop, I was requested to run a night shift on the presses and machines. Now I would venture to suggest that nobody, not even at this time when the country had its fair share of unemployment, welcomed night work, but we readily accepted it as part of the job that we had to undertake whether we welcomed it or not. We all wanted the RT to be a success, a vehicle of which the Board and ourselves could be justly proud.

I must agree that there were certain little strings attached to night work that sweetened up the inconveniences. These were time-and-one-quarter pay for the hours worked, and rather longer hours at that, which made the pay packet look healthier at the end of the week. Up to this time I had never worked on nights, and it was a new experience for me. I had often heard a lot of comments about night work, such as "there is nothing in it", and "you get used to it", usually passed by persons who had never had any personal experience of working at night. However, at 7 p.m. we all clocked in ready to make a start.

The Metal Shop was right across at the other side of the building from the main entrance, so it seemed very strange walking through a factory which we had always known as a noisy hive of industry now still and quiet with not a soul about. It is uncanny in a factory the size of Chiswick Works at night, the only lighting being pilot or

Bus workers as well as buses worked at night. This scene, typical of the late 'thirties shows STL 332, an example of the earlier types of 56-seat STL bus based on petrol-engined AEC Regent chassis with crash gearbox, built in 1933-34. The engines of these vehicles had, in fact, briefly been fitted to LT-type double-deckers built in 1932 which were converted to diesel with new engines in 1933. The vehicle was operating from Willesden garage, where the author had worked in 1921. A characteristic feature of London buses of the 1932-33 period was the shallow arch of the lower saloon ceiling, clearly visible in this view.

police lights, the solitude broken only perchance by a rattle of a chain or some loose gear, and the steam pipes cooling and contracting and thus making a crackle, with an occasional hiss from an airline union.

The foreman had left a note for me containing a list of the various work which had to be undertaken during the night, and I had to allocate the staff to their assigned machines and commence production. They were really a grand crowd of men, and ever so easy to work with. I must add at this juncture that I was the only skilled craftsman on the shift, hence any trouble or complications with the machines, tools, or work produced were my responsibility. The fellows worked hard and diligently, and at any suspicion of trouble they would summon me immediately. We all had a "blow" now and then, and a quick surreptitious smoke, although smoking was still not allowed in the works at this time. Should any press tool give any trouble, I had to strip it down and execute a repair. I often had to proceed to the Tool Room to regrind a tool, then return to the Metal Shop, reset the tool in the press, and continue production. On these expeditions I made sure I was always accompanied by somebody. No breakdown could wait till the morning if we could possibly effect a repair, as this would lose valuable production time.

The machines would be switched off at about midnight for an hour's break. How still and silent it would seem then! We used to sit in a group alongside a bench with just one or two lights overhead switched on, while the rest of the factory was in darkness, like a "small tribe of workers" at an oasis in a desert of gloom. The canteen was closed, hence we all had to bring our own food and make our tea, boiling the water with the aid of the gas blow pipes. I remember one fellow heating up a tin of beans with a blow pipe when, alas, the tin blew up as he had omitted to pierce a hole in it. After our meal, perchance, another quiet smoke, keeping our eyes well open for any intruder of our solitude, such as a patrolling warden or fireman. Making ourselves as comfortable as possible, tongues would loosen and start to wag; yarns would be told and all sorts of subjects discussed. How I have laughed since at some of the tales. Some were obviously true, but others had to be taken with a grain of salt.

It is surprising that when a crowd of fellows, through circumstances, are confined together for a long period how friendly, open, and understanding they can be. The atmosphere of course was free and easy. There were about a dozen of us, and we had a job to do and meant to see it done, but we had nobody watching us and we worked our own way and in our own time. You certainly get to know people better when working nights with them. I remember one fellow used to start the shift by taking a liver pill, before break he would take bicarbonate of soda in a glass of water, during the early morning more powder to settle his stomach, and before going home a good dose of salts. He used to tell me he knew "how to keep fit". However, he was always on the job and never lost time, and it is surprising what the human body can stand.

Thursday night was pay night. My first task was to hand out the pay cards, and when the staff were satisfied that all was correct on the credit side we would all go to the cashiers (it was a late night for them) and draw our cash. It was useless for me to attempt to allocate any jobs until the cash had been received and sorted out, indeed I wouldn't even try! Little did any of us realise then what many of us were to experience in the days that were yet to come, that we should be working nights in this same shop during the early days of World War Two making black-out cowls and headlamp masks for the road services fleet, including the very vehicles that we were helping to make at that time.

Night work to me was an experience of working as a member of a team endeavouring to perform a duty to one's employers under adverse conditions. It gave me a welcome opportunity to get to know my fellow workers, and in some measure to understand the other fellow's problems. It upset one's own domestic and social life and had obvious problems regarding meals. It was a complete reversal of the normal morning-to-night timetable of everyday life. It also taught me to be resourceful should any problems arise where under normal day-work conditions you would rely on the factory procedure and administration. But these were not available at night, and therefore I had to seek my own solution. Looking back now, I am glad I had the opportunity to work nights. The experience gained, I am sure, has been an asset to me in my subsequent career in bus work, and I am confident it never did me any real or lasting harm.

Inspection

The inspection and approval of units after overhaul at the garages in the old days, as described in earlier chapters on Forest Gate and Leyton garages, was carried out by the foreman. He was a real engineer, a man who had served his time as an apprentice in the shops and, to use the term not often heard these days, he "knew the game and the drill". When he decided something was O.K. this was official, and nobody would dare dispute it. The individual components of the units would be passed or rejected by the fitter responsible for overhauling the unit. He would be governed usually by the

This historic photograph, taken inside the Licensing Section at Chiswick in June 1939, shows RT1, the prototype of the RT class, ready to go into service. The complete line-up of vehicles gives some idea of the range of types then represented in the fleet. From left to right, they comprised of a T-type single-decker (one of twelve originally in the fleet of Thomas Tilling Ltd.), a new TF-class Leyland underfloor-engined Green Line coach, RT1, an STL, an LT-class six-wheeled single-decker, a new STL of the final pre-war standard type (with an ST-type behind), a Q-type side-engined country bus and the rear of another Green Line coach. All except the TF, RT and the new STL would have been at Chiswick for overhaul or repair. Note that all but the TF were on AEC chassis.

feel or slop in the mating parts. Nobody worried much about standards, as we understand them today. The main idea was, did the shaft spin properly, and was it true and not too sloppy? Did the bearing bed satisfactorily with no apparent rock? Spare parts were drawn from the stores and fitted by hand, and very little machining was undertaken. The completed vehicle was finally passed after a road test by the garage tester, and driven, bearing trade plates, to the licensing authorities, who would pass or reject it as being fit for public service.

This procedure continued very much the same at the four Central Depots, as also described in an earlier chapter on Willesden garage, but with the opening of Chiswick Works and the subsequent introduction of mass-production methods of overhaul it became obvious that a rigid inspection of individual parts had to be inaugurated. Hence as each chassis part was dismantled, and after passing through the washing machine, it was conveyed to the section known then as the Second-Hand Viewing. Here all components were inspected and tested, and either rejected or passed for further service. After inspection, the detail parts were marked with a small spot of paint to note their condition and destination in the works procedure. The colour code used was green, fit for further use; blue, for repair; and red, for scrap. In later years this coding has been considerably extended to include the following:-

Green	—	Fit for further use
Green/yellow	—	Dress cams, gear teeth, etc.
Green/brown	—	De-plate
Green/black	—	Undercut for nickel plating.
Green/white	—	Undercut for chrome plating.
Green/blue	—	Undercut for welding
Yellow	—	Magnetic test completed
Yellow/pink	—	Change or assemble, plug, stud, etc.
Yellow/white	—	Machine after processing (weld, plate, & metal spray).
Brown	—	Machine

Brown/white	—	Braze
Blue	—	Weld
Blue/brown	—	Undercut for metal spray
Blue/orange	—	Metal spray
Orange	—	Heat treatment
Orange/white	—	Overhaul
White	—	Straighten or rebush
Pink	—	Fit heli-coil insert
Pink/white	—	Machine or bush
Red	—	Scrap

The staff engaged on inspection work in the early days were known as "viewers", not as "inspectors" as they are today. It was unwise for anybody from any other department to hang about the Second-Hand Viewing Section for too long, or else perchance, unknown to you, the back of the heels of your shoes would be given a liberal daub of coding paint. I have seen fellows, employed in the works, on Chiswick Park Station platform homeward bound with painted heels, unknown to them of course.

All new parts from outside traders and manufacturers were inspected in the Main View Room, and had to be certified as satisfactory before being passed into the stores as stock for use in the works or garages. But there was no inspection of Chiswick's own finished parts, such as those machined during overhaul, and it was considered the responsibility of the person doing the job to see that the dimensions, and assembly if any, were correct. Should any queries arise they were referred directly to the foreman concerned.

Early in 1939 an inspection system was commenced on the various sections within the factory. The first phase was the arrival on the various selected sites of large surface plates, with vee blocks, angle plates, and all the other paraphernalia associated with inspection which had never previously been seen outside the confines of the Tool Room and the Main View Room. The appearance of this equipment, as can reasonably be expected, was viewed with some apprehension by the workshop staff, who were wondering how "tight" and rigid the dimension limits were going to be applied, and how this new procedure was going to affect them. There was the usual speculation as to what was the big idea. We all knew the obvious answer, of course, and shop floor inspection had to come, especially as the Board's vehicles were becoming more and more complicated by incorporating new features of modern design and construction, and also the new RT2 double-deck body was now being built in the Metal and ancillary shops.

Inspection staff were recruited from among the skilled craftsmen from the shop floor who showed an aptitude for inspection, and who were quick at the correct reading of micrometers and height gauges and sharp at spotting errors. Successful applicants were regraded as inspectors and given a differential rate of pay above that paid to skilled craftsmen. And so the inspection of all piece parts and units was started, and very soon settled down. Arguments arose of course, as we all knew they would, but these were invariably soon overcome by mutual co-operation between the workers and the inspectors. I remember one particular inspector, only recently promoted, was especially zealous in his position and would not consider passing any machined part unless it was "dead on size". He was promptly and for ever afterwards known as "Spot-on-Harry".

With the introduction of this new inspection routine the responsibility of the foreman over the work produced in his section was taken away from him, which gave him more time for administrative and supervisory duties. In addition, a jig, tool, and gauge inspection room was inaugurated, and a rigid control of all tools was maintained. Completed units were thoroughly tested for performance on test beds or fixtures before their assembly into the chassis or (in later years) into the "A" and "B" frames of Routemaster vehicles. The body structure and electrical equipment was also thoroughly inspected, and the completed vehicle was road tested, prior to being submitted for licensing.

Backed with all my experience of "Bus Work" I am confident that when a bus has been overhauled at the Executive's works and certified by the authorities it is as mechanically and structurally sound as any public service vehicle anywhere in the world. I am sure that the general public do not realise the amount of time and trouble that is taken to ensure that the Board's road service fleet is as safe as human endeavour and thought can possibly make it. Inspection of parts, whether new, secondhand, or reconditioned, is an expensive procedure, but it is essential under modern conditions. It is sometimes described as a costly and necessary evil, a field wherein "non-producers" can dwell, and which harbours many inspectors who are perfectionists and are always far too ready to "snag" other people's work. This is not basically true, of course, and to achieve the very best results from any product, whether mechanical, structural, or even decorative, inspection in some form or other is essential. I sometimes think there are occasions when it can be overdone, by quibbling over minute details which in some instances don't mean a thing, and which costs a considerable amount of time and money and hinders the production figures. However, on the whole a happy medium is usually agreed.

6. Wartime

Preparing for war

Late in 1937 the Board, in common with many other large undertakings, started to form Air-Raid-Precaution squads at Chiswick. Ludicrous though it would seem, only supervisors or staff in administrative positions were invited, or even allowed, to join. On joining this A.R.P. organisation, members attended a series of lectures and demonstrations, and ultimately were issued with protective clothing and equipment. Subsequently monthly exercises in all squad duties were arranged. The workshop staff, other than by news received through the grape vine, were entirely unaware of the existence of any preparations or plans proposed or tentative for Chiswick Works. As is so often the case, the personnel who later were destined to play such an important role in active A.R.P. organisation were the very last to be informed.

During 1938 bus work at Chiswick continued in very much the same pattern as previously, though by this time the Metal Shop had been redesigned to cope with the building of the new RT2-type bodywork, as already mentioned, and the prospect of this work had been received with great satisfaction by the staff concerned. The works holiday taken in the first week of August had come and gone, but world tension was growing day-to-day as war with Hitler seemed increasingly certain. The industrial first-aid squad members were canvassed and requested to be the nucleus of the Chiswick A.R.P. first-aid squads. It was indeed commendable that all these fellows volunteered to a man, especially when previously they had been ignored.

Men from the workshops were now recruited for digging trenches and filling sandbags. While reflecting on seeing these fellows digging the trenches, I can recall the shaped timber gauge that had been made in the works, being dragged about on the site prior to being lowered into the trenches to obtain the correct depth and size. As can well be imagined, very few men engaged on this work had had any previous experience, and they all worked zealously, for nobody knew how soon any of us would be pleased to take shelter in them. The weather was perfect, and the gauge was in constant demand. Nobody could possibly foretell at this time what form the war would take, but use of poison gas was naturally presumed likely.

As early as July 1938, the Board had been approached by the Home Office with a view to planning the conversion of Green Line coaches into ambulances at 24 hours' notice. Designs were subsequently prepared, and a set of equipment was produced in the Metal Shop at Chiswick, which, having been found satisfactory and approved, the Board received instructions on 5th September to proceed with the programme. This order, as can well be imagined, was a top priority job for the Metal Shop, and there was no lack of volunteers to work overtime or night work so as to execute the work in the shortest possible time. In only a fortnight the whole of this equipment had been produced, and distributed to the appropriate garages. A year later, on 2 September, 1939, on the day before war was declared, all Green Line coaches were withdrawn from service and converted into ambulances, and many vehicles continued to be used for non-PSV purposes for the duration of the war although the Green Line services were resumed in 1940 using double-deck buses.

We all breathed a sigh of relief and thankfulness brought about by the late Mr. Neville Chamberlain's appeasement policy, and the respite gained by him at Munich in September 1938. We realised that invaluable time had been gained for Great Britain to prepare herself, at least partially, for the inevitable conflict. The London Passenger Transport Board was no exception to the rest of the great organisations which profited by this interlude. During the winter of 1938 and the early part of 1939, the air-raid shelters which had been dug at Chiswick at the time of the Munich crisis were reinforced and cemented. Seating, lighting, and telephone to the control post were installed, and the whole structures were made very much more solid, and able to withstand considerably more blast and bombing.

Control posts in reinforced concrete, half buried in the ground, were constructed in various locations in the confines of the works boundaries. One of them was situated adjacent to the main gate, one close to the Bollo Lane gate, and another at the rear of the First Aid Surgery. The works telephone exchange, which had been situated on the first floor in the old main office block, was now resited in an underground strong post incorporated in the new 1938 building which nowadays is used as the Recruitment and Training Centre

Meanwhile, the overhaul and maintenance work of buses and coaches continued at Chiswick to the scheduled programme as previously planned, but obviously with these preparations proceeding in the event of war a certain tension was apparent as to what we were all heading for. The works annual holiday had again come and gone when the war clouds began to gather once again. Undeterred by wishful thinking, or by the daily newspaper headlines which announced that there was no possibility of war, we in bus work were not so sure, and neither evidently were the Board, who continued active preparations to be able to cope as far as possible with any emergency which might arise. We all realised that come what may, another Munich would not solve this new crisis which was growing more serious hour by hour. The inevitable happened, and on Sunday 3rd September, 1939, at eleven o'clock a.m. the nation heard the Prime Minister broadcast on the radio that Great Britain was now at war with Nazi Germany.

The ARP squads

An Air-Raid Precaution (A.R.P.) Organisation was set up by the L.P.T.B. just before the outbreak of war in 1939, and I joined this scheme straight away.

I was on A.R.P. night duty on the Saturday just before the declaration of war on the first Sunday morning of September 1939. Everybody had been expecting this for several days, and officialdom was ready with its various emergency measures. I well remember driving my little Austin Seven car to the works that night, without any lights, as instructed on that evening's B.B.C. radio broadcast, along roads devoid of any street lighting, and with all the premises en route completely blacked out. It was a new experience for me, and I was not at all happy about it, and was relieved to arrive at the works safely. In fairness to the Board I must say that we members of their A.R.P. organisation were well equipped to deal with any foreseeable incident. I was appointed as leader of the squad to which I was assigned, and I would here pay tribute to all its members for their loyal support and assistance so spontaneously given throughout the duration of the war. We shared very many experiences together which made us all respect and understand one another, and helped to create and cement a comradeship that will last as long as we all may live.

I was very conscious of the responsibility that I had undertaken to bear, and I would like to place on record the fact that throughout the whole duration of the war I was never criticised or admonished by any member of the administrative staff of the Board over any action which I took regarding any incident which occurred while my Squad was on duty. During the entire war I was indeed honoured and privileged to represent all the first aid personnel on the Works A.R.P. Committee. Meetings of this Committee were convened every few weeks to give the representatives of the Fire, Gas, First Aid, and Spotters Squads an opportunity to raise any requests for new equipment, new procedures, new duty rosters, matters relating to liaison between the squads, and in fact suggestions for anything pertaining to the welfare and efficiency of the organisation.

The chairman of this committee reported directly to the Works Engineer, hence there was no delay if immediate action or sanction was essential. On more than one occasion the Works Engineer attended our meeting to greet us all, and to encourage us in the job we were doing. The function of this committee was well worth while, and I am confident it gave satisfaction to the management and A.R.P. personnel alike.

I remember being on A.R.P. duty on Christmas Night of 1942. Thank heaven we had no alert during our 12-hour spell of duty, and my squad and I spent the time playing cards, drinking many cups of tea — nothing stronger, and smoking wartime "Tenner" cigarettes which we were lucky enough to obtain. These were in short supply and rationed by the works canteen at this time, but they very wisely and considerately reserved some for the A.R.P. squads. Strange to say, by a mere coincidence I was on duty with my squad on the very last night of the war as well as the very first.

Bus work in wartime

The first rude and dramatic effect of the war upon the public was the imposition on 2nd September, 1939, of dimmed lighting, which was known as the "blackout". From that day onwards every vehicle was compelled by law to run with very reduced lighting, which quite naturally led to a general slowing down of traffic. Headlamps were fitted with metal cowls and largely blanked internally by cardboard discs which had a horizontal slot running across the centre. The interior lights of buses were covered with metal cowls, these too having a small slot so as to allow a small beam of light to penetrate the saloon. Much later, metal headlamp masks were manufactured in the metal shop at Chiswick, to a Ministry of Supply design. These were of a cylindrical canister type, with one end fitted to the headlamp by means of the canister back-plate, and the other end having three horizontal slots running across the cowl, with the metal so displaced folded round to serve as three shrouds. As the war progressed, many other designs were tried from time to time for masks and shrouds to black out the lighting on the Board's vehicles. The greater part of all the black-out devices used during the war on the Board's road service vehicles were manufactured in the metal shop at Chiswick Works.

While I was working in the Metal Shop on night work we had a very urgent job carrying out a modification to the headlamp cowls, which had just been approved by the Ministry of Supply. For some unknown reason the dimensions of the aperture in the cowl had to be almost "spot on", although this was a stupid tolerance, much closer than was necessary just for letting a beam of light through. It was probably set by a top-level bureaucrat at the Ministry who more than likely did not know the slightest thing regarding the job. However, my task was to set the tool in the power press so as to produce the desired product correct to drawing. In setting these presses, which had adjustable stops, you first produce what is termed a "try-out". Whilst setting this particular press I had two try-outs, which were both slightly oversize, so I forthwith handed these to the

machine operator standing nearby and gave him definite instructions
to scrap them. After again slightly adjusting the relevant dimension
on the press tool, and re-checking it, and satisfying myself that all
was now precisely correct to the drawing, I then gave the order to
proceed with the main operation.

Later that night, after we had produced many hundreds of cowls,
imagine my surprise to have a visitor, no less a person than the then
Chief Engineer (Road Services) of London Transport. This
gentleman asked me if the job was under control, and he stressed
that the dimension was important. I assured him that all the work
we had produced was correct to drawing, and I picked several up at
random and measured them in front of him for him to see. After a
brief word with me he departed, and I learnt subsequently that he
had forthwith gone to the Paint Spraying Shop, where our jobs were
being sent through the machine and sprayed black. Here he picked
up two of our jobs which had already been sprayed, out of a pile of
many hundreds, but by a thousand-to-one chance or more he had
selected the two "try-outs" that I had instructed to be scrapped.

My word, what a "flap" then followed. All through the rest of
that night an inspector was engaged on the task of trying every single
one for size. I was quite confident that he could not and would not
find any more that were not correct to the drawing, and in due
course I was proved right. This incident is just one that fortunately
was very rare, thank Heaven, but actually in this case it created a
very big scene over such a very minor dimensional error which had
inadvertently slipped through the machining because of one individual not carrying out his instructions. Following this episode I made
perfectly sure that any subsequent "try-outs" were scrapped by myself, either cut into two parts if possible, or else rendered unusable
with the aid of a hammer.

By this time we had already said good-bye to all our fellow
workers who were Territorials or Forces Reservists. Some of us who
had previously had commitments as Forces Reservists but were now
time-experienced or "free" were now considered to be very lucky,
but time alone would tell. It is appropriate here to mention that the
Board and its predecessors had always been very kindly disposed to
any staff who had commitments with the Forces. The winter of
1939 passed slowly into the early spring of 1940. This was what the
Press termed the "phoney war" period, and the Board took every
advantage of it to build up its A.R.P. (Air Raid Precaution) service,
to improve the black-out and air-raid shelters, and generally to
complete the transition of Bus Work from peace to war.

Then in May 1940 the war really started, followed by the collapse
of France, and, alas, Dunkirk. We all realised then that Britain was
now really fighting for survival. At this time the works received an
order to build 20 heavy armoured vehicles. It was a top priority job,

Three STL-type buses and one STD negotiating Piccadilly Circus in wartime. The leading
bus has had one lower-deck window boarded up, probably as a result of blast damage. Roof
panels, hitherto painted silver, had been repainted in an unvarnished red oxide to make
vehicles less conspicuous from the air, while white patches were painted on mudguards to
make them more easily seen from ground level in black-out conditions.

and there was no lack of volunteers to work as we did right round the
clock to execute this order, which must have been completed in
record time. They were mounted on ex-Tilling ST-type bus chassis,
and built of quarter-inch steel plate. The weight of the vehicles must
have been enormous, but they evidently satisfied the ordering
authority, for we in the workshops heard no more of them. Also at
this time there came Anthony Eden's "call to Arms", and the Local
Defence Volunteers organisation was born. Many bus workers
responded to the call, and the works was patrolled by armed men of
the Chiswick Unit, to guard against sabotage and parachute attack.

The almost nightly bombing of London produced an acute shortage of serviceable buses towards the end of 1940 and provincial operators lent some 473 buses of assorted types for service in London. This Leeds City Transport AEC Regent, dating from 1932, was mechanically almost identical with a contemporary STL, but many of the buses were of models that had never hitherto been operated by LPTB. Note the bomb damage in the background of this 1941 photograph.

Dunkirk had another immediate effect, which was very soon apparent at Chiswick. Due to the "call-up" of male staff for service in the Forces, and men being directed to aircraft factories, there were no longer enough men to carry out all the work. So the Board began to recruit women in the unskilled engineering grades, and for the first time in its history Chiswick numbered women among its workers. Women usually started as labourers or cleaners, but were given every encouragement to advance. Some with previous experience were engaged as coil winders in the Magneto & Dynamo Shop. The work the rest of them did was very varied. They cleaned vehicles, swept shop floors, pushed sack-barrows and trucks, and operated power presses, multi-drills, and power punches in the Metal Shop. Some women drove Lister petrol-engined trucks in the confines of the works, while others drove motor-cycles fitted with box sidecars carrying mail to and from 55 Broadway and the various garages.

Enemy air raids started in August 1940, and very soon became the order of the day. The sirens kept on wailing, and the constantly repeated air-raid warnings caused an excessive amount of lost time from the benches and machines, for in those early days of air raids all work was immediately stopped on the first sounding of the "alert", and all personnel took cover in the trenches situated outside the factory buildings and stayed there until the sounding of the "all clear." But it very soon became obvious that if this procedure was allowed to continue the whole war effort would be in jeopardy, and so the "spotter" system was borne. A look-out post was built on the roof of the highest point of the main office block, and manned in 12-hour shifts day and night by qualified men drawn from the factory staff, who gave instant warning when they could see the approach of enemy aircraft. The spotters could broadcast over the works loud-speaker system, and say "Take cover" when at their discretion it was wise and prudent to do so. This system would be more difficult nowadays, on account of all the much-higher buildings that have since been erected in the surrounding district.

From the time of the introduction of the spotter system little or no regard was taken of the ordinary public air-raid sirens, and the staff in Chiswick Works relied solely on the spotters giving adequate warning of impending danger. These fellows did a grand job of work and commanded the respect of all. When the "take cover" was given, all staff proceeded to their allotted shelter as quickly as possible, and all the First Aid, fire, gas, and rescue squads stood by at the ready to meet any emergency. When they considered the danger had passed the spotters gave the "all-clear" to return to work. These were certainly grim and dangerous days for the population of the Metropolis, and those of us engaged in bus work were no exception, for I often thought what a wonderful target from the air Chiswick Works must look.

Gas masks and tin helmets were carried by everybody, and were always kept near at hand ready for use. In all my long experience of Bus Work I have never known such comradeship and understanding for the other fellow as existed at that time. The common danger seemed to unite us all, and the petty differences of everyday life were swept away in the grim struggle, when no one knew whether they would see the end, or in fact what the ultimate end would be.

It was very soon learned that the best protection against blast for the Board's rolling-stock was cotton netting glued to the inside of all glass windows. A small diamond-shaped aperture was cut out of this netting in the centre for passengers to view their whereabouts. When windows had once been blasted and broken they were boarded up, and in most cases remained so for the duration of the war. Protection of buildings against blast was provided in the first place by sand bags, later to be superseded by specially built blast walls and the bricking-up of some windows.

As the progress of the war dragged on, some of the work from Chiswick was decentralised to various garages. Although this resulted in some inefficiency, and a lot of extra effort in carrying things backwards and forwards, it was intended as a security measure, as well as enabling nearly half the factory to be used for aircraft production.

By the end of 1940 so many of London's buses had been "blitzed", some beyond repair, that an S.O.S. for help had to be sent to the Ministry of War Transport. This was answered by the despatch from 51 provincial bus undertakings of 473 buses on loan to help. What a galaxy of strange vehicles they were! They came from everywhere from Cornwall to Inverness, all painted in varied liveries and unusual colours, of nine different chassis makes and 24 different chassis types, many of which had never before been seen in London, not to mention the multiplicity of body types, and including oil-engined, petrol-engined, single-deck, and double-deck types. And what a headache they were for us to maintain! Unfamiliar vehicles, for which in most cases we had no spare parts, meant that these had to be made in the works, copying from a sample which had by now become worn-out or broken.

At this time nobody worried too much about the "paper work" for each job, which is such an important feature of any job to be undertaken today. Suffice it to say that the shortage of spare parts and materials even for our own standard-type vehicles grew acute, so the recovery and repair of worn or damaged parts had to receive the most careful attention. Many materials in short supply had to be replaced by substitutes, and nothing was scrapped that could be repaired and given another lease of life, or used for some other purpose.

As the war progressed, surface and trench air-raid shelters were constructed in the confines of the various workshops, primarily to afford closer protection for the night-shift workers, who also served in the role of "fire watchers". In those days it was quite common, especially for toolmakers or key men, if working on a priority job, to be asked to work on and complete the job if you can. This often meant working for 24 hours, sometimes even longer,

During the later war years, substantial numbers of 'utility' buses were supplied to London Transport. The majority—some 435—were on Guy Arab chassis, but there were 181 Daimlers, mostly CWA6 models, of which the first six, supplied in 1944, were fitted with lowbridge bodies—a layout not previously used among the red buses of central London. These were allocated to Merton garage for route 127, and D6 is seen here as delivered, with wartime anti-blast netting on the window glass. The diamond-shaped openings allowed very limited vision, but frustrated passengers tended to scratch at the netting to gain a better view of the scene outside, despite the increased risk of the glass shattering if a bomb landed nearby. The bodywork was built by Duple.

except for meal breaks, and to work late in the evening and then have to walk home while an air-raid was in progress was quite a regular occurrence.

I still have very vivid memories of coming into the works one Monday morning, after a week-end of very heavy air-raids, to the toolmakers' bench in the Metal Shop, where I was employed at that time. We found a bomb had fallen in the factory in the near vicinity of our location the previous night. Glass, black-out, and fragments of asbestos roofing were scattered everywhere, and we had no lighting or power for the machines. But in a matter of only hours, by the united and untiring efforts of all the staff, and by calling on all available resources, the debris was cleared up, the power switched on, the machines working, and the roof covered again. What an achievement! It was typical of the prevailing spirit among the workers that became so apparent in an incident such as this.

The "Music While You Work" programme, from the B.B.C. sound broadcasts, which is nowadays regularly relayed morning and afternoon over the works loudspeaker system, was introduced in the early days of 1940. The strains of the music was supposed to give a fillip to the factory staff for them to concentrate greater personal effort and achieve maximum production. In some measures I think it did boost production, indeed it was generally recognised by the management to be a worthwhile feature of works routine, and has been regularly continued from then up to the present time.

At long last V.E. Day dawned, for Victory in Europe, and after nearly six long years of war we at least in Europe were at peace, soon to be followed by V.J. Day and Victory over Japan, to complete the peace for which we had dreamed and longed for so long. The lights began to shine again minus their black-out shrouds, and then began the long and gradual change for peace-time. The works A.R.P. organisation was disbanded, and all the personal equipment issued for wartime was handed in, and later despatched to a central depot, where most of the useful items were resold to members of the staff who wanted them. I regret to relate that the A.R.P. personnel, many of whom served since the inception of the team, received no official thanks, either verbal or written, from the management, and likewise neither were their individual staff records held by the Board suitably endorsed to record this service. To leave such occasions without any official thanks did nothing to foster good relationship between management and staff.

Blood donors and Mass X-rays

Soon after the start of the Second World War many of the staff at Chiswick Works responded to the very worthy call from the Emergency Blood Transfusion Service for the donation of some of their blood, which at that time was very urgently needed. Notices would be posted on the works notice boards stating that the mobile unit of the Blood Transfusion Service would be in attendance on the premises on such-and-such a date, and inviting members of the staff to volunteer to donate some blood. If they were thus willing they were asked to give their names to the Sister in charge of the works First-Aid Station or to a member of her staff, who would later inform them at what time they were required to attend. If the number of prospective donors was somewhat lower than required the Sister or one of her nurses would go into the offices or work-shops and make at random a personal appeal to certain members of the staff. Needless to say, by using this tactic they always got more donors than they really required.

I remember I got myself in her "bad books" once, when I suggested to the Sister and her nurses that they should set an example and put their names onto the list before ever they approached anyone else to do likewise. Did they "go off" at me! "Because we are all so vulnerable to infection in our work", they replied. Nonsense, I told them in no uncertain manner. I think that the fellows in the workshops run a far greater risk of infection every day of their lives than any person working in such a wonderfully clean and well-equipped first-aid surgery such as that in Chiswick Works. It has always been a very sore point with me to hear, as I have so often heard in my long career in bus work, that this or that job is "Dead easy", or that there is "nothing in it", usually spoken by a person with very little knowledge of the task or the problems associated with it. My idea has always been that one should set an example, and have a go, and do all the talking (if you must) afterwards. Don't ask other people to do something that you would not think yourself of doing, or at least of making an attempt to do it.

But to return to the Blood Transfusion Unit. This would arrive at the Works, and accommodation would be provided by the management, and the various donors would arrive at pre-arranged times. A doctor and a nurse would be sent with the unit, and one of our own First-Aid men would also be in attendance. He would be there to give "on the spot" assistance should it be required, and also to act as an orderly. I was indeed honoured at one day's session to be asked if I would attend for this purpose. I readily agreed to this, and spent a very worthwhile day, and I am sure learnt quite a lot about people's behaviour and reaction regarding the donation of blood. Before starting, the doctor would ascertain the donor's blood group, of which there are four, and he would be in attendance the whole time.

After donating their blood the patients would be given some hot sweetened tea, and allowed to rest until they felt well enough to resume their normal job of work. A few of them passed out, in a faint, but they usually came round again in only a few minutes. During any one day some two hundred or more people would have attended the venture. Our management and staff co-operated very well indeed with those associated with the unit, and I never heard one single complaint from anyone.

Another similar event which occurred on about four occasions, a little later on, in the sixties, was a visit to the works by the Mass Radiography Mobile Chest "X"-Ray Unit which was operated by the North-West Metroplitan Regional Hospital Board. This was a four-wheeled Leyland Beaver lorry chassis with high and very square bodywork, pulling a four-wheeled trailer of the same shape and size, both being painted plain white all over. One unit contained accommodation for the removal of much of our clothing, especially anything metallic, and the other one contained the X-Ray camera and the card index records. I have vague memories of some kind of canvas connection being rigged up between the two.

The Unit was installed inside the confines of the works building, in the north-western corner, and operated by staff from the Hospital Board. All our own staff were invited to submit their names and working locations to the Personnel Officer if they wished to attend, and they were subsequently told of the day and time they were to present themselves. A very high percentage of the staff agreed to take advantage of this opportunity, including almost the whole of the Drawing Office staff, of which by now I was one. I took the view that should I have any chest abnormality that required investigation the sooner it was revealed and some action was taken the better, and this conception was probably held by the vast majority who attended.

The process took only a few moments, and we all filed through in a more-or-less continuous stream. We had to stand with our chest up against a vertical plate, which was in line with the camera lens, the operator adjusted the height of a little bracket on which we had to rest our chin, then he clicked the shutter, and a serial number on each exposure on the film in the camera coincided with that on a card on which our name and other details were written, and that was that. In due course we all received a card sent to our home address some days later, and I suspect all heaved a sigh of thankfulness, as I certainly did, to learn that we were "all clear". But I was sorry to learn later that unfortunately a few of our fellow bus workers were found to be in need of urgent further investigation and treatment. This outcome made it abundantly clear to us all that this first-ever exercise to be held at Chiswick Works (in March 1960) was indeed well worth all the organisation, effort, and expense involved. Further visits of the unit followed in February 1963, January 1966, and March 1969.

The war years allowed valuable operating experience to be built up with the first production batch of 150 RT-type double-deckers which had entered service between November 1939 and June 1940. Here RT 132 receives a wash in Victoria garage in May 1942. The bodywork on these vehicles, classified as type RT2, was the last major batch of bodywork to be built at Chiswick but its design provided the basis for the post-war RT3 and subsequent types of London Transport design built by Park Royal and Weymann. Note the rear route number box, characteristic of the pre-war RT designs but eliminated on post-war examples.

7. Post-war progress

The RT family of buses

In 1945, soon after the end of World War Two, the RT3 type of double-deck bodywork, an improved version of the pre-war Chiswick-designed-and-built RT2, was designed in the Chiswick drawing office. It was envisaged that this would ultimately constitute the great majority of the post-war double-deck fleet in operation. Orders were placed for these new bodies to be built by the Park Royal and Weymann coachbuilding firms, also for the corresponding 3RT chassis to be made by the A.E.C. factory at Southall. Delivery, from both firms, was being made from May 1947

The war years, with much of Chiswick Work's efforts diverted to war production, inevitably resulted in a lowering of maintenance standards, quite apart from the effects of wartime overloading and in some cases bomb damage. Replacement of old vehicles was well behind schedule and bodywork had to be repaired by makeshift means to keep vehicles on the road until sufficient new buses could be built. This photograph of STL 1430, built in 1939 and hence due for replacement by pre-war LPTB standards in 1946-47, was running in post-war livery in March 1950. Note the use of heavy gauge external strapping to reinforce three of the pillars.

onwards. Further orders were placed late in 1947 from these and other firms (Saunders, Craven, Metro-Cammell, and Leyland for bodywork, also Leyland for RTL and RTW chassis officially coded 7RT and 6RT respectively), bringing the total at that time for which contracts had been placed to over 4,000. Eventually, over the next few years until 1954, some 6,805 vehicles of RT and similar types were built and operated in service. This included 500 of the increased width (RTW) type, which at that time had only quite recently been sanctioned by the Minister of Transport, and to run only on certain specified routes. This restriction would seem futile in the light of present-day bus operation, when it is realised that 8 ft.-wide Routemaster and Daimler Fleetline vehicles are now operated on a very large number of routes.

Unlike the 2RT-type vehicle, the chassis frame of the 3RT finished immediately behind the rear axle spring shackles, so that it did not support the rear-entrance platform, this being cantilevered from the main body structure. The delivery and subsequent operation in service of the new 3RT vehicles was very welcome both to the Board and the passengers alike. The war years had taken their toll of the existing fleet, and the appearance on the roads of brand-new buses, not to be confused with wartime utility ones, was indeed a pleasing sight; public transport was gradually returning to and indeed improving on normal pre-war standards of comfort.

It should be recorded that London Transport no longer built the bodywork for the majority of its new buses, as had been the case until World War Two and indeed had been LGOC's practice since horse bus days. Chiswick Works had a huge backlog of body overhaul from the war years, with many buses still in service that would normally have been scrapped, and many requiring body repair to keep them on the road. It was considered best to make arrangements for Park Royal and Weymann to build the Chiswick-designed bodies for the RT family of buses.

Lofting for accuracy

On 28 August, 1945, the management advertised on the time-recording clocks for applications from suitable personnel to fill some temporary vacancies in the Drawing Office at Chiswick. The successful candidates would be engaged on "lofting" work in connection with the production drawings required for the building of the new post-war double-deck RT3 bus bodywork. This was the first vacancy circular for administrative positions that had ever been

The appearance of the first post-war RT-type buses was doubly welcome because of the improvements they brought for both passengers and crew and as much-needed reinforcements for the London Transport fleet. The first vehicles, RT 152 with Park Royal body and RT 402 with Weymann body, appeared in May 1947, and production was at first fairly slow, with about a dozen buses from each bodybuilder by August. RT 159 is seen when brand new on arrival from Park Royal at Chiswick. Note the early type of route number display on the front lower-deck bulkhead pillar—this was intended for ease of identification when the bus was too close to read roof-mounted numbers.

addressed to the works staff in the whole history of the works, though it was the forerunner of what is now a recognised procedure when vacancies occur. As can well be imagined, the circular was received with apprehension, and with some suspicion, as are all new ventures and procedures when introduced into workshops.

However, I felt very confident that I would be capable of such a task, and decided to make a written application. I was lucky enough to be granted an interview, and, after attending, was requested to take a practical test on the work required to be undertaken. Having attended for the test, I evidently satisfied the panel of my ability

Some idea of the volume and variety of overhaul and repair work in hand at Chiswick in early post-war days is conveyed by this view of the paint shop in November 1947, with D86, a Daimler CWA6 with Duple body, and the Park Royal body from a wartime Guy as well as an RT2 body of the 1939-40 period receiving attention. Many of the wartime vehicles, as well as being non-standard, tended to suffer from the lack of suitable materials for their construction, with inadequately seasoned poor-quality timber requiring extensive repair at an early age. These reasons added to the urgency of rapid production of the post-war RT.

and skill as regards "lofting" work, and was later offered a temporary position on the technical staff. Consequently, on 12th October, I started my career in the Drawing Office, after being employed on bus work in many workshops for nearly 27 years, and thus achieved my long-cherished ambition.

For the uninitiated I must give some explanation as to what "lofting" means. It is the process of making extremely accurate full-scale drawings, usually of large assemblies and contours. It originated in shipbuilding, and the term "lofting" was derived from the fact that, owing to the space required to produce the large-size drawings of a thing as big as a ship, it was often accomplished in the lofts of buildings. Messrs. Handley Page, Limited, the well-known aircraft manufacturers, introduced lofting into their design offices during the Second World War, and certain principals of the Board came into close contact with it here, because Chiswick Works, together with

four manufacturers in the transport industry in the London area, jointly produced all the components required for no fewer than 710 Handley-Page Halifax bombers. This enterprise was known as London Aircraft Production. Having thus gained experience of aircraft manufacturing methods London Transport decided to adapt these to bus work.

At this time the Board had already placed large contracts for the building of new RT3 bus body with Messrs. Weymann Limited and with Park Royal Vehicles Limited. The Board wanted these buses to have the precise accuracy and complete inter-changeability and standardisation of components which Handley-Page and other aircraft manufacturers were already achieving, and hence lofting was introduced at Chiswick. Complete assemblies and sections were drawn full-size with special silver-solder-pointed instruments on to painted aluminium sheets of size 4ft. by 8ft. These sheets did not

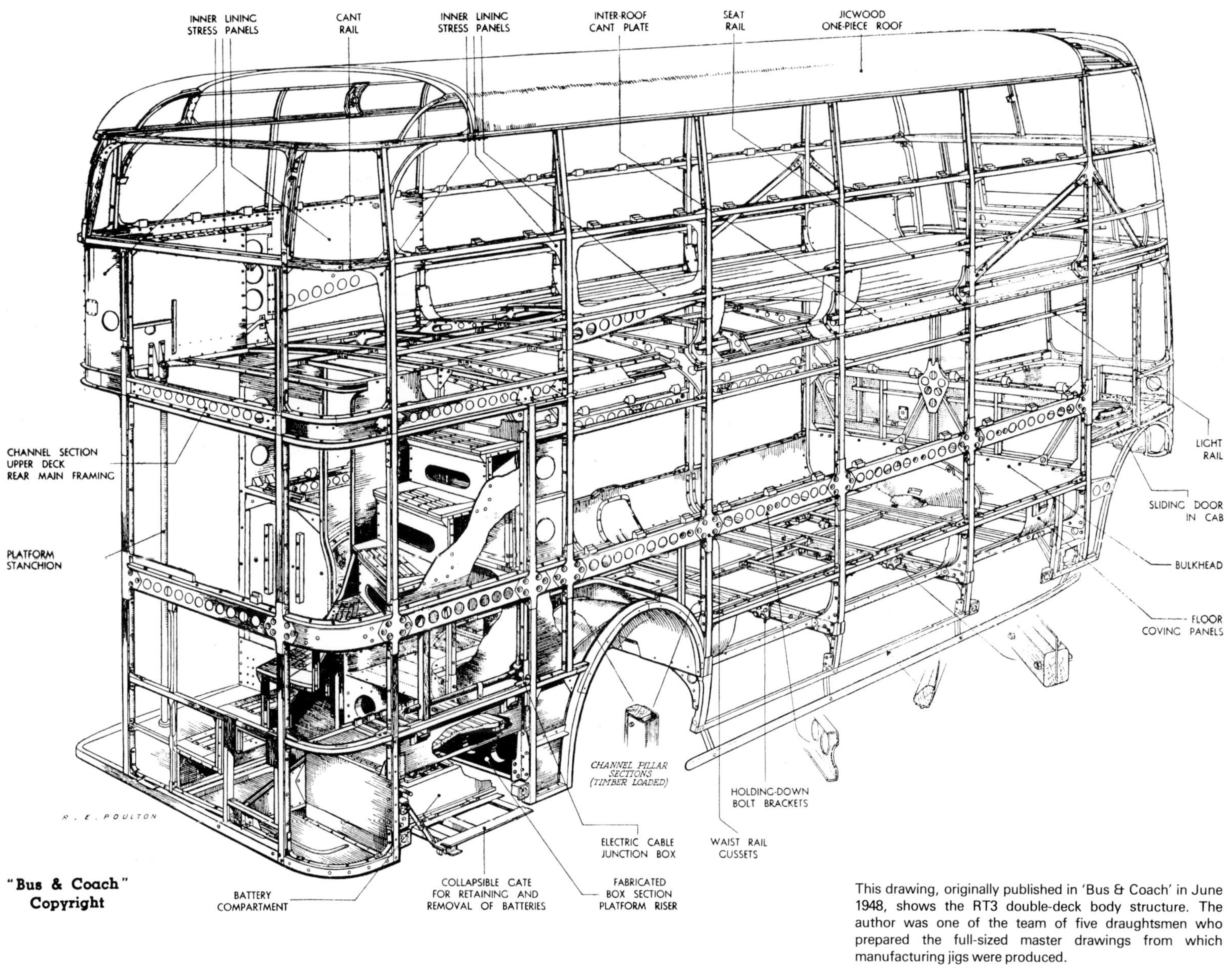

This drawing, originally published in 'Bus & Coach' in June 1948, shows the RT3 double-deck body structure. The author was one of the team of five draughtsmen who prepared the full-sized master drawings from which manufacturing jigs were produced.

99

Dimensional accuracy extended to the fit of chassis with bodywork. Here a Leyland chassis for an RTL bus is checked in one of the special alignment jigs to which every RT and RTL chassis had to conform, identical jigs being provided at the AEC and Leyland works.

The close family resemblance between AEC and Leyland buses of the basic RT design is emphasised in this view showing AEC RT 925 with Leyland RTW 402, the latter being one of the 500 vehicles with Leyland-built bodywork of 8ft. overall width instead of the 7ft. 6in. of other London buses of that time. The Leyland body and chassis conformed to RT standards in other key dimensions, appearance and type of transmission but incorporated Leyland engines and axles.

stretch or shrink with the humidity of the atmosphere in the way that paper does, and hence the dimensions remained exactly the same as when drawn. These lofted drawings were then photographically reproduced for the manufacture of jigs, tools, and templates, which in turn were used for the mass-production of the bodywork components.

This work was very interesting, and called for an extremely fine degree of accuracy. The developed lengths and shapes of panels and brackets had to be calculated, and drawn on to the sheets ready to be photographically reproduced for the bodywork contractors. The lofting sheets were marked up in 12in. squares, with lines which were called "grid lines". These were scribed on to the sheets through the medium of a grid table which was fitted with locating pins to determine that the lines were scribed accurately. The purpose of these grid lines was to minimise any inaccuracy in our drawings, because we always took our measurements from the nearest grid line and not from the far end of the sheet. All the lofting tables and the grid-line marking tables were manufactured in the works tool room. The lofting section of the Drawing Office became quite a show piece to visitors while it remained, but on completion of the preparation of the drawings and the necessary templates required for the manufacture of jigs and tools for the building of the RT3 bus bodies, the section was disbanded, and has never yet been revived. The five draughtsmen employed there, including myself, and all five of whom had come up from other parts of the factory were then found alternative work elsewhere in the Drawing Office, where all of them remained for many years.

Incentive bonuses and new staff

During 1946 an incentive bonus scheme was inaugurated in the works. As first introduced the scheme was tentative, but it was hoped that it could be placed on a permanent basis as soon as practicable. The introduction of the bonus scheme made it necessary to have a Planning and Ratefixing Office, to control the routing of all jobs through the factory. It was also essential that the detail parts of certain components and sub-assemblies should be more fully drawn and dimensioned than previously, and this requirement applied particularly to alterations and modifications in design that had been made to older components.

This system of payment by results was soon found to be very encouraging both to the Board and to the staff, whose earning capacity was greatly improved, and a greater output and efficiency resulted. It seems strange now to recall that prior to the inception of this scheme the presentation of any bonus scheme by the management had been most vigorously opposed by the representatives of the works staff. Right from the inception of the new bonus scheme the supervisory staff have received a percentage of the works average rate, but at no time have any of the administrative staff, either clerical or technical, participated in any bonus scheme.

All new staff when first appointed, from the early fifties onwards, have been given a copy of a booklet entitled "Welcome to London Transport". This was an extremely good publication right from the start, giving copious information to the newcomer and a very good insight into the staff policy of the undertaking. New entrants are also given background talks designed to provide them with a comprehensive knowledge of the activities and organisation of London Transport early in their careers, and especially of the aspects that apply to themselves and to the department or section where they are employed.

London Transport Executive, 1948

On 1st January, 1948, in accordance with the provisions of the Transport Act of 1947, the undertaking of the London Passenger Transport Board was vested into the British Transport Commission, and its management was delegated to the new London Transport Executive. The responsibility of the Board for the provision of passenger transport in the London Passenger Transport Area thus ceased on 31st December, 1947, and after that date the Board continued in existence only for the purpose of winding up. It was exactly fourteen and a half years since the Board had commenced operations,

RT production was just getting into its stride as the London Transport Executive came into operation on the 1st January 1948. This photograph of the 100th post-war RT to be delivered, RT 441, leaving Chiswick after preparation for operation from Potters Bar garage, was taken in November 1947. Within eighteen months the total was well over 1,000.

on 1st July, 1933, having been formed six weeks previously on 18th May, 1933.

The conclusion of the Board's existence marked the end of an era, and the beginning of a new one with the London Transport Executive taking its place as one of the main constituents of the national transport system. To the passenger the transfer was carried out with no noticeable difference in his daily means of travel. To the personnel at Chiswick Works this change of control would have passed completely unnoticed had it not been for the amendments to correspondence headings and the titles of documents, also for the need to change the transfers denoting the legal title of the operating authority on the near-side of all vehicles. At this time one of my numerous jobs in the drawing office was to design and amend all the transfers for lettering and figures displayed anywhere on or in the road services vehicles. Consequently I had to prepare a revised drawing to cater for this change of ownership, to show a new transfer with new lettering. These new transfers were duly printed and delivered, but most of them were affixed to the vehicles overnight in the various bus garages, in itself quite an achievement, and we at Chiswick had to change the name only on those vehicles which just happened to be in for overhaul, or for any other reason, at that particular time.

Aldenham Works

As far back as 1920 the L.G.O.C. had decided to centralize the overhaul of their fleet at Chiswick, in order to improve the efficiency of the work, reducing the time the vehicle was off service by having the whole process of overhauling under one roof, and thus reducing to a minimum the transportation of units. After the end of World War Two it was envisaged that the then rapidly expanding road service fleet would ultimately reach the 10,000 mark. In anticipation of this it was decided by the management that the overhaul requirements of such a fleet could not be entirely undertaken at Chiswick Works, and so it was decided to build a new factory, or to find suitable premises on a site of sufficient size within the London area.

Eventually it was decided to build the factory on 54 acres of land at Aldenham, near Elstree, in Hertfordshire, using as a nucleus an existing factory which had been built in 1939 for a railway depot to form part of a scheme (which never materialised) to extend London Transport's Northern Line from Edgware to Bushey Heath. This building was fully occupied all through the war in assembling 710 Handley-Page Halifax bombers, the individual components for which had been made in five other factories owned by five organisations working jointly under the name of the London Aircraft Production Group. (One of these five was London Transport's Chiswick Works.) Conversion of Aldenham Works from railway to bus requirements commenced in 1952 and was completed in 1956. It was formally opened on behalf of the Minister of Transport and Civil Aviation on 30 October, 1956, and was then heralded as the largest and most up-to-date plant in the world for the overhaul of road passenger vehicles, equipped with the very latest flow-line production methods. It is ironical when one reflects that this factory is built on land at one time most rigorously controlled and zoned as part of London's Green Belt.

On the opening of Aldenham Works vehicle overhaul work was divided between Chiswick and Aldenham as follows. Allocated to Aldenham was the overhaul of the vehicle as a whole, both body-work and chassis frame, the manufacture of body parts, both timber and metal, and the repair of accident damage to vehicles which was too complicated and extensive to be dealt with by the staff available at the operating garages. Allocated to Chiswick was the overhaul of all the mechanical and electrical units and their detail parts. This scheme of dividing the work between two factories was a complete reversal of the policy which had been born with the opening of Chiswick in 1921. It is extremely hard for the layman in bus work to understand the merit of operating two factories more than 13 miles apart, with the enormous expense of transport involved in the

(Opposite page) Perhaps the most dramatic aspect of Aldenham Works is the overhead travelling gantry crane and its ability to lift a double-deck bus body high enough to clear others of the same type when being moved to or from the repair bays. In this view, taken shortly after the Works came into operation in 1956, an RT series body is lifted over an RTL on which the body has been mounted before painting.

(Above) The high bay in the body shop has an almost cathedral-like quality, such are its lofty proportions. In this early view, RT double-deckers and the underfloor-engined RF single-deckers of the same period look almost like Dinky Toy models.

The body inverter makes it a simple matter to work on the underside of a double-deck bus body. Here an RT body is steam-cleaned.

The precision-built structure of the RT body greatly simplified repair, as parts and sub-assemblies could be directly replaced without the need for making replacement parts to suit appreciable variations in dimensions as had been necessary with traditional bodybuilding methods.

division of work between the two of them. But, alas, the total size of the bus fleet did not grow as large as was expected, and so during 1967, after only 11 years, the overhaul programme had contracted so much that about one-third of the Aldenham factory was leased to Leyland Motors Ltd. as a spare-parts and service depot, and as a result of this the timber sawmill has now been relocated back at Chiswick again.

In the writer's opinion, the three outstanding features most worthy of note at Aldenham, are:- Firstly, the overhead travelling crane which lifts a complete bus body high up above other bodies and moves it to the body repair shop, where it is placed on stilts which are so sited as to be in a desirable working position, and so dimensioned as to be identical with the body locating points on the chassis frame, and are set at a convenient height for work, so that the body is thus held on what is in reality an adjustable jig conforming to the appropriate type of chassis. Secondly, the body inverter, in which the body is held and can be tilted to an angle of up to 90°, giving complete access to the underside in an extremely ingenious

manner, which allows many functions hitherto arduous and costly to be completed underneath the body easily and quickly. The third feature of outstanding merit, never previously used on London's buses and coaches, is the paint-spraying booth capable of containing three vehicles and producing their exterior livery, with its heated oven for drying newly-varnished vehicles.

It is readily agreed that many functions previously carried out at Chiswick have, with a more modern layout and advanced equipment, been greatly improved. The whole problem of workshop technique has been studied throughout the entire layout of the factory. Special gantries have been provided to facilitate access to the exterior of the vehicle, without the need to use portable steps and ladders. It is a modern factory, light, spacious, and ideal for fresh air, but in an out-of-the-way location for staff access by public transport. London Transport has to operate special staff buses for the employees, on more than a dozen regular routes, morning and evening, from and to most of the western, southern, and eastern suburbs of London.

The Routemaster

In an undertaking such as London Transport development cannot stand still, so an entirely new design of vehicle was developed, with many new features never before used in bus design. The first prototype was shown at the 1954 Commercial Motor Show, and three other prototypes were built in 1956-57, all four running in passenger service. The first production model was shown at the September 1958 show, with a somewhat different frontal appearance from the prototypes, and a number of detail differences made as a result of operational experience with the first four. The RM, or Routemaster, vehicle was designed for increased seating capacity, minimum weight, and ease of control, servicing, and overhaul. Most of them seat 64 passengers, and are 27ft. 6in. long, as compared with 56 seats and 26ft. 0in. for the RT and STL types. All the RM's are 8ft. wide instead of 7ft. 6in., but there had previously been two batches of eight-footers in London, i.e. the 500 Leyland RTWs from 1949 and the 43 South African trolleybuses at Ilford from 1941 onwards.

With a wheelbase of 16ft. 10in., the Routemaster has no conventional chassis frame. The body structure, built by Park Royal Vehicles on all production vehicles, serves as the main load-carrying unit, with two sub-frames to contain the mechanical units at front

The prototype Routemaster, RM1, in the form in which it was revealed to the Press in February 1955, having been built at Chiswick Works, using AEC mechanical units. The radiator was mounted under the lower saloon floor, just behind the offside front wheel, and hence the front panel carried a bullseye motif rather than a radiator grille. This was done to reduce the bonnet length and hence give more room in the lower saloon within the 27ft. overall length limit which then applied. Note the provision of only one forward-facing destination indicator and a small panel over the platform—both were changed before the bus entered service, which did not occur until February 1956.

RM2 entered service in the Country Area green livery on route 406 from Reigate garage in May 1957 but later that year was repainted red and transferred to the Central Area garage at Turnham Green, which frequently operated experimental vehicles because of its proximity to Chiswick Works. By this time the length limit had been increased and the radiator reverted to its conventional position. This bus was also built at Chiswick, though with some participation by Park Royal, and incorporated AEC mechanical units. It incorporated numerous minor changes from RM1.

Two of the key mechanical features of the Routemaster are evident in this view of the front subframe of one of the two experimental vehicles with Leyland units (RML3 and CRL4). The front subframe was attached to the integral body structure and carried the independent front suspension,(a feature which is still uncommon on buses over twenty years later),as well as the engine, steering gear, etc. These vehicles, with Weymann and Eastern Coach Works body structures respectively—each to LTE specification and appearance—entered service in 1958 and 1957 respectively, CRL4 being a Green Line double-deck coach.

for the Unified threads were completed and published by the British Standards Institution in July 1949 as British Standard No. 1768.

Due to the almost complete absence of timber very few wood screws are used, and there is a total absence of the welding which had hitherto played such an important role in bus construction. The gangway, floor, and platform are covered with a rubber-cork composition, and bevelled slats and tiles of this material are fixed on by a waterproof adhesive, covering the whole area of both floors, as a departure from the previous standard practice of fitting timber slats, lino, and metal tread plate.

A notable new feature on the Routemasters, known to all of us on bus work but probably never noticed by the passengers, is the stainless steel stanchions, handrails, and seat frames, which have superseded the long succession of those previously made in aluminium alloy. The feature which the passengers probably most appreciate is the heating and ventilating system, designed and developed by London Transport. The volume of water passing into

and rear, these being supplied by A.E.C. and easily removable for overhaul. Coil springing is fitted at both front and rear, the front suspension being independent, this being entirely new to London Transport. Steering, although the conventional worm-and-nut type, incorporates hydraulic ram-type power assistance. The four-bay body structure was fabricated from high-duty aluminium alloy throughout. The front bulkhead carries the driver's cab, front cowling, bonnet, and nearside wing, while the rear platform and staircase are suspended from the upper-saloon the same as on the RT3 design of 12 years previously. Special attention was paid to the need for ease of replacement of front and rear-end body parts, by the use of H-section extruded aluminium-alloy pillars. Corrugated aluminium-alloy sheeting performs the load-carrying functions of the upper and lower-saloon floors, with plain sheet of the same material for ceilings and floor coverings. Hardened threaded steel bushes are pressed into the exterior flanges of the pillars and rails to accept stainless-steel screws for fixing the exterior panels, and solid and tubular aluminium-alloy rivets are extensively used throughout the body structure. The management decided to embody, both for the body and the mechanical units of the Routemaster, the then new Unified screw threads in place of the British Standard Fine and Whitworth threads used hitherto. Specifications

Production of Routemasters was entrusted to Park Royal, using AEC running gear, and RM73 conveys the appearance of the earlier production vehicles built in 1959. Although conventional and, by 1979 standards, of a layout rendered obsolete by the subsequent general adoption of one-man operation, the Routemaster was very advanced in structural and mechanical design—many of its features still being ahead of conventional practice twenty years later.

Early deliveries of Routemasters were allocated to trolleybus replacement duties. This is the scene in Poplar garage in October 1959, with a new fleet of Routemasters ready to take over from the trolleybuses just visible on the left of the picture.

both saloons is regulated by a control in the driver's cab which operates a blind passing between the exterior grille and the heat exchanger. Apart from this control, the volume of air flowing into the vehicle depends on the external air velocity and the vehicle speed. Eleven complete air changes per hour can be achieved with the speed of 25 m.p.h. In cold weather the air is heated as it passes through a heat exchanger under the front destination box. This radiator is inter-connected with the main radiator and engine cooling system, and thus uses some of the heat from the engine. A heat control is located near the top of the centre of the lower-saloon front bulkhead. This varies the flow of water through the heat exchanger, increasing from no flow for cool air to full flow for warm air, so that a saloon temperature of 25° above ambient is obtainable.

In my opinion the Routemaster is a first-class bus for the conditions of operation prevailing in London and suburbs. The orthodox positions for most of the main features have generally been followed, but in the main have been greatly improved. I feel some features have been retained because the London Transport fleet has always had them, and that others, such as blind lettering and displays, have been altered with little or no advantage to the passenger. The passenger appeal aspect of the design and livery has been rather overdone. What the passengers require is a reliable service to take them from A to B in the shortest possible time. They are far more inclined to notice a dirty vehicle, and to comment on it than the colour scheme that has been applied. It is worthy of note that the traditional red exterior livery, associated for so many decades with London's buses, has been maintained.

It seems to me that far too much time and expense was spent on modifications which in the main mean little if any advantage to the passengers, and serve only to depart from having a fleet of standard vehicles to maintain. Even slight departures from standard design are costly, especially where an overhaul procedure embodying a complicated bonus system is employed. I would suggest that all the features of these modifications should be noted and embodied in any future design. Those involving safety, premature failure, or where considerable saving on costs results, would automatically be embodied on existing vehicles.

The cramped position of the nearside and offside lower-saloon front seats has been retained on the RM, and the flywheel cowl encroaches into the lower saloon just as much as on every previous type, proving an awkward obstacle to encounter when about to use or leave these front seats. The longitudinal foot-stools in the rear of the lower saloon are much improved, but the problem of the differential cowl is more pronounced than ever, remaining a potential stumbling block to both young and old. The conductor's locker and full-length coat compartment, however, are well designed and desirable.

But still, despite all these little criticisms, I still think the Routemaster is a very good bus, and I am very proud to have been personally associated with it, as one of the team in the drawing office which helped to design it. I produced well over 100 drawings for the manufacture of the Routemaster, ranging from such features as the flywheel cowl, foot-stools, rear section of the bottom frame, interior finishers, and bulkhead blinds, down to the transfers and a host of varied detail components. Towards the end of the production run a lengthened version appeared, being 30ft. long with 72 seats. The structure and components of this are exactly the same as the 64-seat version, except that the whole bus is, in effect, cut in half in the middle, and an extra section or bay, somewhat narrower than the other four, is inserted.

London Transport Board, 1963

The Transport Act of 1962 provided for another reorganisation of the whole nationalised transport system, this time into a number of separate Boards, corresponding to the previous divisions of the British Transport Commission. The broad effect of this new Act to us engaged on bus work in London was the transfer to a new London Transport Board as from 1st January, 1963, of the facilities previously operated and the responsibilities previously exercised by the London Transport Executive. To the general public this transfer was carried out almost unnoticed, just the same as with the 1948 reorganisation. There were no big headlines in the national Press to herald the birth of this new Authority, and certainly there were no outward signs of any change in the transport undertaking of the Metropolis and its suburbs.

To the personnel at Chiswick Works this change of control in 1963 would, just as in 1948, have passed completely unnoticed had it not been necessary to make arrangements for the garages to change the near-side exterior transfer on all road services vehicles denoting the title of the operating authority. Likewise, correspondence headings and titles were suitably amended. To most of us in bus work this was considered merely a "paper change", and made no real difference in vehicle maintenance or overhaul. It is easily revealed how little the general public know of the operating authority of public service transport in the London area, when reference is occasionally made in the Press, even now, to the London Passenger Transport Board as being responsible, when in fact this body died in 1947.

Yet another change, reviving the 1948 title "London Transport Executive", was made in 1971, and here again, this could be regarded as another "paper change", for although there was an important new relationship with the Greater London Council, the day to day aspects of bus work were not affected. However this did not occur until after my retirement, but it was noticeable that the changes of the lettering on the vehicles in 1963 and 1971 were not carried out with the promptitude that had applied on 1 January 1948.

All become "Mr" (but some still have nicknames)

Until January 1963 it was an instruction from the management that all correspondence to officers and principal executives was to be addressed to "Esq". This procedure, associated with the Victorian era, was hopelessly out of date, and the Board are to be congratulated that this instruction was rescinded. Henceforth all male staff, irrespective of position, were to be addressed as "Mr". To the author this change seemed long overdue, considering that up to this time all correspondence to the workshop staff had not even been given the title of "Mr", but merely a name coupled with a clock number. This lack of common courtesy did not conduce to good relationships between management and staff, and the issue of this new instruction, while seeming very trivial, can now really help and be of great value in fostering better harmony in industrial problems.

Nicknames are commonly used in most, if not all, sections of the community throughout the length and breadth of the country. Bus work is no exception to this, and many nicknames are in constant everyday use in the offices and workshops. Some are given in affection, some in derision, some relating to a peculiarity in the stature or make-up of the individual, or to his place of birth, or perchance relating in some measure to kind of job in which he or she is employed. I remember a fellow in the garage where I started my career in bus work, who was nicknamed "Springy", the reason being that he was the man who examined, and changed if necessary, all the springs on the vehicles operating from this garage. Wherever the workshop there is invariably to be found a "Tom Thumb", "Big Bill", "Nobby" (Clark), "Noisy Ned", "Lightning", "Chalky" (White), or other similar names which are self explanatory.

Sometimes surnames are shortened or clipped, such as "Sam" for Sampson, or "Jack" for Jackson, and yet, peculiar as it may seem, some other names are lengthened, such as "Smithy", "Brownie", "Shortie", etc. Moreover, as can be expected, there are always the Jocks, Taffys, and Paddys. Even the highly respected sister-in-charge of the first-aid surgery at Chiswick was known by all the workshop staff as "Iodine Kate". She knew this, but it caused her no offence, though of course nobody would ever dream of calling her this to her face; Heaven help them if they dared to! Even the worthy executives of the Board were not immune from being labelled with a nickname, and, to mention a few renowned to all Chiswick workers, there were "Tiger", "Blue Eyes", "Auntie", and "Donald Duck".

In the author's opinion the origin of a nickname is incidental, and the use of such names in either workshop or office does no harm. If it fosters harmony and good relationship between people who are endeavouring to pursue a good and useful job of work in any sphere of bus work then by all means let us keep them.

In my own particular case, I also have always been known by a nickname. Right from birth, even at home, I was never called by my proper Christian name of Henry. I was always known by a nickname, though I won't tell you what! When I started working at Forest Gate garage there was at that time a well-known boxer fighting under the name of Bob Scanlan, though whether he was of national or only local fame I cannot now remember. I had the same surname, so somebody at the garage promptly called me "Bob", and I was quite happy when this name "stuck" to me. I never did like the name "Henry", and I always considered it as having a rather simple ring, and certainly not suitable or indeed acceptable to me when starting career in bus work. And so in every garage, workshop, or office where I was subsequently employed I was always known as "Bob" to everybody, and I still am today.

Hours and holidays

Over the period covered by this book it is very interesting to reflect on the gradual reduction of hours worked in the normal working week, also the introduction of paid holidays and how they have been extended from time to time. When I started in 1918 the working week was 54 hours, but in 1919 this was reduced to 47. This remained static for 30 years, when it was reduced to 44 hours. In 1960 this was further reduced to 42 hours, and in 1965 to only 40. This gradual reduction of working hours, of course, was not peculiar only to bus work, but applied generally to all large industrial concerns. This was part of the normal trend in industry as a whole,

and the management were not too slow in conforming with this fairly wide-spread policy which was gradually being applied throughout the country.

Paid holidays, or leave with pay to the workshop staff of the company, were totally unknown until 1927. During that year three days' leave on full basic rate of pay was granted. This was considered by one and all as a really wonderful gesture from the company, and until then some employees had never known the joy and satisfaction that leave granted without loss of wages could mean. This was certainly a step in the right direction, and was increased in the following year to four days, and in 1929 a full week's paid holiday was granted. In 1931 the management announced that in future Good Friday and Christmas Day would also be paid holidays. In 1947 the introduction of a fortnight of paid holiday, plus six Bank Holidays, first took place in the works. The management must be congratulated for bringing this into effect, for I am sure it was long overdue, and I am certain its introduction has paid good dividends and has been much appreciated by the staff concerned.

The five-day week, which is almost universal in industry today, became operative in Chiswick Works from late in 1947, and from that date production work has not been undertaken on Saturdays. The three-quarters-of-an-hour lunch break, instead of a full hour, and finishing a quarter of an hour earlier in the evening, was introduced purely as a temporary measure in 1940 during the dark days of World War Two, but in fact it has now been retained as a permanent feature. Its retention seems ludicrous if one reflects for a moment on the bitter struggle and experiences suffered by workers in the distant past to obtain an hour's lunch break. The 40-hour working week was introduced for all workshop staff as and from 17 March, 1965, having been reduced in stages from 54 in 1918.

But throughout this period the working hours for all administrative staff remained static at 38, though to be fair it must be recognised that the Board have been very liberal with regard to annual and statutory holiday entitlement for all grades of administrative staff, and over these years many improvements have been implemented. However, the five-day 35-hour week came into operation from Monday 16th September 1974, several years after I retired, and it now applies to all London Transport's Office staff in the administrative, technical, professional, and clerical grades. This is the first change in office staff hours since before the First World War, and although many welcome improvements in hours had been made for the factory staff the "white-collar" workers had had to mark time until now.

8. Modern methods

Research and development

In the very early days of the opening of Chiswick Works in 1921 the facilities for the development of "bus work" were very limited by comparison with the standards when I retired in 1968. A very small Drawing Office was situated at the western end of the north-wing office block, but the entire staff, including the draughtsmen, an apprentice from the works, two men on the specification section, and a lad covering general duties and printing, was never more than a dozen including the Chief as well. Most of the drawings were produced on paper, very few on linen, and prints of these were taken on a very old and antiquated vertical carbon arc-lamp machine. The lamps traversed from top to bottom, and the drawings were fed into the machine from the side, while ancillary equipment for developing and washing was located close by. All prints in those days were "blues", hence the term blue-print. These prints would fade if left too long in exposed light, so the drill when using them in the work-shops or offices was to lay them face down when not being looked at.

Development of mechanical parts and units was undertaken in the Experimental Shop, which included a very small laboratory not much larger than the dining-room of an ordinary suburban house. The original laboratory was mainly concerned with the testing of petrol and lubricating oils, but over the years that followed the need for scientific research increased, and the scope of the department was widened. In 1931 new laboratory premises were constructed in the works, in a location now housing the Personnel Office, and work was undertaken for other departments as well as bus work. The provision of up-to-date research facilities is economically essential in such a large undertaking as London Transport. Very few of the passengers carried in the Board's rolling stock realise the amount of technical and scientific research that has taken place, even on the materials that are required to construct the vehicles, or the fuel that they use.

The amount of research work grew so extensively that it was agreed to extend activities into the railway field, and to have constructed a purpose-built laboratory specially designed to serve the needs of the Board's railways as well as its buses. So a new two-storey building with an adjacent single-storey annexe was built at Chiswick Works alongside the Broad Street railway line, on a site which had previously been the seasoning store for timber. In addition to equipment transferred from the previous premises many new machines and pieces of apparatus were installed. Metallurgical, physics, and engineering activities are housed on the ground floor, while the chemical sections, the library, and general administration are on the first floor. This new building was officially opened on 5 December, 1960, by Mr. A.B.B. Valentine, the then chairman of London Transport Executive.

During this period while the laboratory research work had expanded, so likewise the work in the mechanical development field also increased. In 1930 a separate Development Office was inaugurated, and the Drawing Office, while working in close liaison, was rehoused in different accommodation, in a long thin strip along the southern edge of the main factory building. At the end of World War Two the Drawing Office was again relocated, this time in a first-floor office inside the western end of the main factory, in a portion of the factory which was itself an extension of about 1937 and whose roof was somewhat higher than the rest of the main building. This office, making good use of this higher roof, had been built in about 1941 when the whole of that end of the factory was leased to London Aircraft Production. At that time it had served as a general administrative centre, with a stores on the ground floor underneath it.

The bus Drawing Office moved in to here in 1945, but it very soon had to expand considerably so as to cater for designing the new post-war RT3 bus body, so the maintenance section of the drawing office (dealing with alterations to older types of bodies, and also chassis and engines) moved out to The "Hut", leaving the RT3 bodywork people (and there was by now a very large crowd of them, including the lofters and the specification section as well as the draughtsmen) in sole possession of the ex- L.A.P. office. The "Hut" was a portable wartime wooden building which had been erected outside the main part of the factory, exactly opposite the top end of the main drive up from the front gate. I personally worked in the "Hut", on bodywork maintenance drawings, for about eight years, and it was quite warm and cosy despite its temporary nature. In February 1953 the trolleybus drawing office was transferred from Fulwell Works to Chiswick, and was located in the "Hut" with us. There were only five of them, but we had to squeeze up a bit and let them have one end, because they still functioned as a separate self-contained unit.

In 1955-56 a very substantially built new extension was added to the main administrative office building, as a new south wing, on a site that had previously been the staff car park. In July 1956 the entire Drawing Office staff, both from the "Hut" and from the fomer L.A.P. offices, was transferred to the first floor of this new building, and the staff of the Rolling Stock Engineer, who had latterly been in another wooden hut, near the canteen, moved into the ground floor. This new Drawing Office, in which I worked for my last 12¾ years, is exceedingly light and spacious, certainly the best

Exterior and interior of the 'Hut', in which the author worked on bodywork maintenance drawings for about eight years up to July 1956.

The framework of the new Drawing Office in course of construction in 1955. The author worked in this building for the last twelve and a half years of his career.

Among the projects which emerged from the new Drawing Office were the 30ft.-long versions of the Routemaster which first went into service in 1965. Here one is seen leaving the forecourt of Victoria garage on route 76, while one of the 50 experimental Leyland Atlanteans of class XA approaches on route 24. After a few months, the allocations for these two routes were exchanged to provide comparative operating experience. Some 500 30ft. Routemaster buses were built and classified RML, the last letter standing for 'long' rather than 'Leyland' as on RML3, which was reclassified RM3 to avoid confusion. The Routemaster series of numbers reached 2750, including all variations of the type.

office in which the D.O. staff have ever been housed. At the rear there is a suitably screened and partitioned area housing the print store. A covered bridge has been built to gain access to the first floor of the adjacent wing of the old building, which now houses the specification section, and, to the rear, the print and photostat rooms.

The new Drawing Office has accommodation for up to 40 draughtsmen and four tracers, and four individual offices for executives. It was in this building that the Routemaster vehicle was designed, and all the necessary production drawings required by the building contractors were produced. All drawing boards are fitted with drafting machines, and ample layout space is provided for each draughtsman.

Drawings of preliminary schemes are produced on tracing or detail paper, and final production drawings on linen, with layouts on cartridge paper. Drawing sizes vary from the standard 14in x 10in to 52in x 30in, and for special jobs a size 91in x 30in is available, which is twice the size of a standard drawing board but is very useful for complicated electrical circuits covering a very large area. Most drawings are done in pencil, with dimensions and arrow-heads in indian ink. Tracings are always drawn in ink, and usually the explanatory notes, titles, and part numbers are produced with Uno stencils. The general layout and style of almost all drawings conform to British Standard 308 of 1964, for "Engineering Drawing Practice". This is important, as a very large number of drawings are sent to manufacturers and contractors, and abbreviations of general engineering terms and conventional representation of features and materials are, or should be, clearly understood by all engaged in development of bus work.

The writer, who had the privilege of working as a draughtsman for well over 13 years, has no hesitation in claiming that the high standard of drawings produced in our Road Services Drawing Office would compare favourably with those from any similar industrial organisation. When thinking in terms of drawings, I smile even now when I recall working for a certain foreman many years ago as a young apprentice. He would always draw the job to be accomplished on the palm of his left hand, and when he had explained the job he rubbed his hands and away disappeared the drawing.

Adjoining the Drawing Office is sited the specification section, which is sub-divided into the body, mechanical, and electrical sections. These three play a very important role in maintaining present vehicles, and preparing parts lists and schedules for new ones, and supplying any or all of the details specified in a contract. All modifications to units, sub-assemblies, and detail parts are prepared here, and the necessary machinery is put into action to implement these changes. Any variations in design, specification of material, and size range of contractors' products are noted, and investigated to ascertain what problems there are, if any, and how the changes would affect London Transport.

Very few people except those engaged on bus development realise the legal restrictions that are imposed on the designers. These are for the most part essential for the safety of passengers. These Acts, Orders, and Regulations affect road traffic and vehicle construction throughout the United Kingdom, and control such features as lighting, carrying capacity, equipment and use, standing passengers, conditions of fitness, and weight and dimensions of vehicles. These requirements are extended, amended, and revised from time to time, and often additions to them can mean a tremendous amount of expensive work for London Transport to implement when they become law. The majority, however, are enforced only on new vehicles and from a specified date. London Transport and its predecessors have always been conscious of the necessity of conducting experiments and improvements in the whole realm of bus work, and have made and are continuing to make a very valuable contribution in the design and operation of public service vehicles.

When need arises, London Transport development engineers make use of the facilities at the Motor Industry Research Association's proving ground at Lindley, near Nuneaton. Here the prototype rear-engined Routemaster, FRM1, is seen after climbing the 1 in 6 test hill and about to descend the 1 in 5 gradient for brake testing. This particular photograph was taken during the 'Bus & Coach' road test of this vehicle carried out in June 1967. Although this final development of the Routemaster design never went into production, many of its features have been incorporated in the Leyland Titan rear-engined double-decker now coming into London Transport service.

It is essential in development work to have adequate workshop facilities to produce any equipment or components necessary to conduct experiments or obtain technical data. The Board and its predecessors have always been aware of this, and right from the birth of Chiswick Works there has always been an experimental shop. The building which houses the present Shop was built in 1937 to house the mechanical section. The bodywork section at that time was housed within the main part of the coach factory. In 1946 the two sections were combined under one roof, and the building converted and extended, to become a modern and really first-class and well-equipped establishment.

It is staffed by skilled craftsmen, all specialists in their own trade or craft, who enjoy an appropriate differential rate of pay above ordinary craftsmen. These men are invariably recruited from within the works, and are usually ex-apprentices of the Board. The mechanical section includes an engine test bed and ancillary equipment, vehicle inspection pits, and a variety of machines capable of producing the diversity of parts required for new design vehicles or the development of units or equipment. The body, sheet metal, and electrical sections are similarly well equipped to play their respec-

An indication of London Transport's maintenance standards was given by the decision to refurbish many of the RF class of AEC Regal Mark IV Green Line coaches dating from 1952 when they were sixteen years old and had mostly covered over one million miles each. The estimated mileage of RF89 when this photograph was taken at Chiswick Works in the summer of 1968 was 1,140,000 yet Alan Townsin reported in a 'Bus & Coach' article that it handled as if it was about six months old.

tive roles in the projects undertaken. It is with a conscious sense of pride that the writer records that the first two prototype Routemaster double-deck buses were developed and built in this shop under the combined direction of the development and drawing offices.

Much valuable development work was undertaken in this shop during World War Two, in connection with gas producers for bus operation, also tank engines for the War Office were developed here at this period. Full-size mock-ups of vehicle sections are built here and studied, for example a driver's cab to obtain the best passenger entrances or sliding doors. In more recent years Passimeter equipment and coin-changing machines have been fitted in various positions, studied, and subjected to the most rigid tests, to ascertain the best possible operational location. All types of experimental buses and coaches are road tested by the staff of this shop, and technical data and performance are recorded for examination by the technical officers.

The Board take advantage of the facilities made available to them at the Fighting Vehicle Research and Development Establishment (F.V.R.D.E.) test track at Chobham, Surrey, and at the Motor Industry Research Association (M.I.R.A.) proving ground at Lindley, near Nuneaton, Warwickshire. The two prototype Routemasters were subjected to the most stringent tests at these places. It is often essential in development work to have first-hand records of service data of the component, material, or design feature, which is under consideration for incorporation in vehicle designs or modifications. Numerous experimental features are being tried, proved, and tested by the Executive at any given time, prior to their adoption or rejection. Experiments are conducted by the development office over an extremely varied range of units, components, or materials, indeed anything from, say, a new type of brake liner to a windscreen washer or even a locking sealant.

Some experimental features are fitted to selected service vehicles, operating on different routes with contrasting service conditions. Technical data is then collated at intervals, until at the completion of a set period of time or of mileage a conclusion can be drawn as to what merit or value the item or features have for London Transport. All reports and technical data are filed for reference, should they be needed at some future date. Other features, by reason of their design, necessitate tests to be conducted in the workshops on a test rig. Experiments can result from correspondence or other contact from public organisations and authorities, traders and the general public, and from all grades of London Transport staff, for new methods of construction and design or the introduction of new materials.

The whole development organisation of the road services works is a dedicated team, truly "back-room-boys" of L.T., comprising the technical staff, draughtsmen, and craftsmen, all combining their efforts and skill with the aim of providing and developing the very best possible fleet of public road service vehicles to serve the peoples of the Metropolis. The writer, who has completed half a century of "bus work" divided roughly 50-50 between the workshop floor (mechanical engineering in all its aspects) and the office desk (drawing and pen pushing), would claim without any fear of contradiction that London buses are designed, developed, and maintained second to none anywhere in the world.

The original Red Arrow single-decker fleet consisted of six experimental vehicles with AEC underfloor-rear-engined chassis designated Merlin by London Transport but basically similar to the larger-engined version of the AEC Swift built for other operators. Bodywork was by Strachans. They had XMS fleet numbers.

Standee and one-man-operated buses

London Transport, like many other bus undertakings in Britain, has generally favoured the double-deck bus, because it gives a high seating capacity within compact dimensions and is therefore economical in road space. Short and highly manoeuvrable vehicles are essential, and the open rear entrance on the older type of double-decker gives a roomy one-step platform which allows for passengers to get on and off quickly. It was always considered that a conductor was necessary to issue tickets, give change, check for over-riding, and supervise loading and unloading in order to reduce the time at stops and so ease traffic congestion. Although five standing passengers in a vehicle have been allowed for many decades, when all seats are occupied, during specified periods in the peak or rush hours, it has always been considered that standing in a vehicle was a poor ride, and only undertaken if time was important, or the chance of a seat a few stops further on could be relied upon. It has always been my own opinion that standing passengers were not getting value for money for the distance travelled and the fare paid, this being a disagreeable necessity brought about by time and circumstances.

For many years London Transport operated single-deck one-man buses in country areas, and a few on the fringe of the central area, where the traffic is less intense, the tickets being issued by the driver. But the popular conception of bus operation in the Metropolis itself was completely altered by the advent of London's Bus Reshaping Plan in 1965. On 18 April, 1966, the first-ever standee bus to be operated in Central London commenced service from Victoria Station Forecourt, on a short route to and from Marble Arch which was extended at certain times around a circular route servicing Oxford Street, with a uniform flat fare of sixpence throughout. This enterprise, which was later to be the forerunner of many other such feeder and inter-station services, was named the "Red Arrow". It was advertised as a frequent service for shoppers to the Oxford Street stores, and was very well received by the public as well as being a good financial proposition for the Board.

The service was operated by 36 ft. long single-deck vehicles known then as the XMS class (later MB). These were rear-engined A.E.C. Merlin chassis fitted with bodies built by Strachans, generally to their own basic structural design, but with certain modifications to London Transport's requirements. They were one-man operated with front entrance and centre exit, and powered folding doors operated by the driver, with accommodation for 25 seated and 48 standing passengers. The introduction of these first six Red Arrow buses was to be the forerunner of many others to follow later.

During 1967 an order was placed for 150, and in 1968 for a further 450, single-deck one-man buses of a similar design, some for additional Red Arrow routes, some for ordinary Central Area suburban routes, and some for Country Area routes. These were the MB class, again on A.E.C. Merlin chassis but bodied by Metro-Cammell-Weymann, and also mainly to that concern's standard structural design but to London Transport specification in most

The production Merlin single-deckers had bodywork built by Metro-Cammell and incorporating more London Transport features than the experimental batch. A total of 650 were produced; with several variations of body design to suit different types of service. MB130 is seen in Chiswick Works in July 1968—this was one of the majority of the first 150 that were re-registered before entering service in September 1968, the registration number SMM 130F shown being replaced by VLW 130G. The RF single-decker in the background, also bodied by Metro-Cammell, was then about fifteen years old, but most of the Merlins were destined to have a short life, partly because of reliability problems, but partly because the 36ft. length was found unwieldy in London operating conditions—all but those required for Red Arrow operation had gone by 1976.

other respects, and introducing features which experience with the first six (together with nine similar Strachans "ordinary" buses) had proved to be desirable. Provision was made in the body structure that a conversion from one to another of the three bodywork functions mentioned above could be carried out at the lowest cost and with the least possible amount of work.

The design covering the mounting of the body to the chassis, and all services in this connection, is so arranged that the body can be removed quite easily during overhaul. The chassis frame, which runs the full length of the vehicle, is of bolted construction, the rear end forming a cradle to house the power unit. This is a six-cylinder horizontal 11.3 litre diesel engine, fitted with an 18" diameter fluid flywheel. The epicyclic gearbox is fully automatic and electro-pneumatically operated, having four forward speeds and one reverse. Suspension throughout is once again by semi-elliptic leaf springs, and steering is power assisted. The compressed-air footbrake system has separate circuits for front and rear axles with a mechanical handbrake on the rear wheels only.

The body is of metal-framed non-integral construction with separate entrance and exit, and electro-pneumatically-operated

jacknife doors. Most of the framing is of mild steel, though the partition structure, wheel arches, waist and seat rails and panelling are all of aluminium alloy. Flooring is of 5/8" thick resin-bonded plywood, covered with Treadmaster Multislats and tiles. The seat frames are tubular, while the cushion and squab fillings, also their trimming materials and style, are as those fitted to Routemaster buses. Saloon lighting is by fluorescent tubes, an innovation for London buses, and a heating and ventilating system with automatic temperature control is fitted.

I have described these vehicles in more detail than most earlier types, because working in the drawing office I was closely involved with many aspects of their design. They are what is known in "bus work" as "off the shelf", i.e. more or less a contractor's standard design modified to London Transport's requirements. It seems to me a great pity that all the development work undertaken in producing the Routemaster vehicle, which at that time (1956) was claimed to be the bus of the future, has been totally disregarded. I agree the passenger appeal features have been retained, but now here we were in 1968 back again to the conventional chassis frame, leaf springs, and body structure mainly built of steel. These vehicles are ideal in layout for standee and one-man operation, where an underfloor or rear engine is essential to allow the driver to be able to conduct as well as to drive. This, of course, cannot normally be achieved with the engine at the front where a bulkhead segregates the driver from the passenger.

A criticism of these vehicles is that the driver has no means of knowing exactly how many standing passengers are being carried, once the bus is more than say half full. To me it seems abundantly clear that the orthodox double-decker operated by a driver and conductor, to which London and suburbs have been accustomed for many decades, cannot altogether be superseded by one-man vehicles. One-man buses, are becoming almost universal in most circumstances, as they are already in most other countries. This policy is being forced on us in very many cases by sheer economics.

The Executive has devoted much time and effort to the possibility of developing sophisticated equipment for the automatic collection of fares on one-man buses where a single flat fare is not considered appropriate. However, I am sure the double-deck crew-operated bus will not entirely disappear from London, and there will always be a need for some of them on certain routes in the Executive's area. The one-man operated double-decker has been developed after I retired, and over 2000 of them are now in service including Daimler Fleetline, as well as the new Leyland Titan and Metrobus types. Nevertheless many modern rear-engined double-deckers are in fact being operated with a two-man crew on quite a number of trunk routes.

Long-service awards

In 1958 the management inaugurated a scheme of presenting gold watches to employees who had completed 45 years of service with the undertaking. This spontaneous gesture from the L.T.E. was received with great satisfaction by all, especially by those who at least had a sporting chance to qualify. The idea of presenting a watch to an employee while he is still working is, in the author's opinion, an excellent thing, and I am sure many a recipient must in some measure be proud to wear his watch while still serving the organisation which gave it to him. On the other hand, presenting a watch to an employee on his or her retirement seems illogical, for his or her timekeeping days are now over! After a lifetime of having to be on time they surely want now to relax, and to forget this ever-paramount feature of everyday working life.

All employees, irrespective of grade or position held, having completed 45 years of service in the case of men, or 40 years for women, are now given the option of having a gold pocket or wrist watch, suitably inscribed, or a clock. Or, if they desire to forgo the gift, they can have a cash payment in lieu, but this can only be received when they ultimately retire from the Board's service. Very few employees refuse the gift of a watch or clock to receive the cash award on retirement instead, and it is rather surprising to find that it is usually the higher-salaried staff who prefer the cash.

Watches are presented on behalf of the Board by the employing officer in whose department the recipient is employed. These presentations are usually quite informal, and conducted in a very cordial atmosphere. An exchange of past experiences and events usually follows, and I am sure that this gesture from the Board is well and truly appreciated. Employees retiring after 40 years'service, but without completing 45 years, receive a long service award on leaving.

When I was within a month or two of completing my 45 years of service I was asked by the personnel assistant if I would like to receive my long service award now or to wait until I had completed 50 years. It was pointed out to me that the award for the longer term would be of a greater value, but I elected to receive a watch for the shorter term. I was away on leave when my 45th anniversary arrived, but as soon as I returned I was summoned to the office of the Chief Draughtsman, who accompanied me to my employing officer. I knew both these two gentlemen well, as my work had often brought me into close contact with them. I had not always seen eye-to-eye with them, and we had had our differences of opinion, but I respected them, and I am sure they respected me for my work, ability, and experience of "bus work" embracing so many different fields. I had never been nervous or ill-at-ease when meeting people in higher positions than myself, and so we exchanged greetings, and then followed a very friendly and informal chat. This meeting gave me an opportunity to speak freely and openly to my two "bosses", and in due course I was presented with a gold watch, which is inscribed thus:-

LONDON TRANSPORT
H.D. SCANLAN
IN APPRECIATION OF
45 YEARS' SERVICE

I was very pleased to receive this watch, and well satisfied with the manner in which it was presented. It is a very good watch, and of pleasing design. I am proud to wear it, and it has already given me many years of excellent service. Unlike myself, I recall one fellow I had known in the works for very many years telling me that when he was told he would be presented on the following day with his long-service award he felt so nervous and apprehensive that he did not get a wink of sleep the night before. He had never previously spoken to his employing officer, and had only known him as "The Boss" when he had seen him in the factory. The presentation was an ordeal for him, and he was thankful when the meeting was concluded.

After my employing officer had presented me with my watch he did what I understand is usual on such occasions, and said, "I have asked you a lot of questions; have you anything now that you want to ask me?" Of course, I knew that the "proper" answer should have been "No", but the Chief Draughtsman gasped when I replied, "Yes, I do have something which I would very much like to say to you two gentlemen, and as you have asked me this seems a good time to tell you." I then related to them how I had worked hard and diligently on the Routemaster brochure (as explained in the fourth paragraph of my Introduction to this present book), reminding them that I had been trained as an engineer and a draughtsman, and not as a writer, and yet since the completion of the brochure I had at no time received one single word of thanks or congratulations. My officer, gentleman that he was, rose to his feet and apologised for the oversight and expressed his regret.

I went on to tell both of them that I was now engaged in writing my own book, and my purpose in doing so now was that I would prefer this information to come from my own recollections and not from anybody else. I felt a lot better after this, after I had spoken my mind. I am confident that both my "bosses" must have seen my point of view, and in some measure agreed with me. But I should like to make it abundantly clear to my readers now, that both prior to and following this episode I always worked in complete harmony with both these two gentlemen right up until the day I said good-bye to bus work .

50 years on

Half a century of work completed

On Thursday 8th August, 1968, I had completed half a century of service with London Transport and its predecessors. I had half expected that some small sign of recognition would possibly come from the management to mark this occasion, for, after all, it is not an everyday occurrence, even for London Transport employing a staff of about 75,000. My immediate chief was gracious enough to shake hands and congratulate me, but he would not have known had I not mentioned the fact some time previously.

I would have been delighted to have received some spontaneous and tangible sign from the management to inform me that the day had not passed unnoticed by them. When one reflects that before World War Two all staff in the Board's service at every Christmas were sent an individually addressed greetings card signed by the chairman, in comparison with this so few cards or appropriate messages would be required to send to staff when they had completed 50 years of service.

I am conscious of the fact that those of us who have served the Board for so long are lucky, and indeed fortunate to have achieved this distinction. I am told that when a member of the staff retires with 50 years or more service to his credit he is privileged to have audience with the Chairman of the Board. This interview, in my opinion, should be arranged as near as possible to the anniversary of the starting day, and not postponed until the person concerned has retired and is only too often considered to be a "has been". However, I suppose to some extent the long-service awards presented after 45 years of service remove the need for another function after 50 years, with the advantage that rather more people can perhaps benefit by them.

Retirement from bus work

We often hear in these days that retirement and getting old comes to us all. But this is not true! Only the lucky ones among us retire. Many, unfortunately, pass away before they reach retirement, and, to use the common phrase, they "die in harness". When you are nearing retirement you are constantly being asked by your friends and colleagues, "How long have you to go now?" Some are prompted to ask with a real sense of interest, but I am certain in my own mind that others enquire for the sole purpose of finding out how much longer they will have to "put up" with you, and others with an eye on the position you hold in the organisation when it becomes vacant.

Some idea of the scale and centralised nature of 'bus work' towards the end of the author's career is conveyed in this view in the unit dispatch stores at Chiswick. Taken in September 1960, it shows an AEC engine for an RT being put on a trolley for despatch and Leyland engines for RTL or RTW buses in the foreground together with an AEC petrol engine (possibly for a former bus then still in non-psv use) just beyond them. Preselective gearboxes are prominent in the centre right of the picture.

Staff employed on "bus work", in common with most other industrial undertakings, are normally required to retire from the service of the Board on attaining the age of 65 years in the case of men, or 60 years in the case of women. Members of the administrative staff may retire, if they so desire, on attaining the age of 60 years in the case of men or 55 years in the case of women, or at any time thereafter until the age for compulsory retirement is reached.

It is very surprising the outlook that many employees have on approaching retiring age. For some it cannot come too soon, or so they say. Many others, and these are by far the majority, are full of apprehension as to what this new phase in the autumn of their lives holds for them. In these modern days it is hard for many to realise that many people engaged in "Bus Work" have few, if any, hobbies or other outside interests, and the prospect of long days spent in enforced idleness with little or no interest fills them with gloom and despondency. To these I would comment, get a part-time job, or at least find an absorbing interest in something, preferably entailing some light physical effort and exercise.

The occasion of retirement from London Transport is generally marked by the office in whose department the retiring employee has worked assembling the entire personnel of the office concerned, also all those interested from any other departments, and presenting a parting gift which has been subscribed to by all, together with a long-service certificate (if qualified) from London Transport.

We all realise that this procedure has been built up over many decades as traditional, but unfortunately these ceremonies seldom ring true, and I suggest they should be discontinued. If the staff with whom the retiring employee has worked feel they must give him or her a farewell present, let the chief of the department give it to them in the privacy of his own office with no fuss, and so dispense with all the often largely insincere talk which is frequently an embarrassment to many.

My last day at work

On April 2nd, 1969, I was summoned to the Personnel Office, and informed in a very nice and friendly way that as I would reach the age of 65 on April 7th, that in accordance with the general practice operated with London Transport and indeed most other industrial undertakings my service with them would terminate on Friday, April 11th. Of course, all this came as no surprise, and was fully expected. Realizing that my time in service was fast drawing to a close I had made several announcements from time to time to my various colleagues that I would shortly be retiring. Now I had a firm date to give to all enquiries. I now started to clear out some of my personal possessions that had served me so well during my long service in the Drawing Office. Various text books and instruments which I would see would be of no further use to me I gave away to younger members of the staff. Other books belonged to the Board, and those I placed in order and made sure none was missing. I started to take home the tools and instruments that I had used over the years in both workshop and office. It surprised me to find how much equipment and jumble I had accumulated. Some I had bought that I had used very little, while other items had given me very many years of valuable service.

At last the final day of April 11th dawned. I left home at the usual time, so as to be at the office at least ten minutes before the official starting time. Game to the last, I was not going to break my long-standing rule and arrive late on this my last day. I thought what a contrast this was from my starting day 50 years ago! Then I had been a small lad as described in the beginning of this book, and now I was an ageing man with a lifetime of experience of all the many changes that had taken place in bus work. I was now dressed in a smart suit, with hair cut short-back-and-sides, complete with a hat and carrying an umbrella and brief case, as had been my custom for a quarter of a century. I had over the years often been chivied about always wearing a hat and carrying an umbrella, and had sometimes been nicknamed "the umbrella man". But alas, this life style is fast declining, and is being replaced by a "sloppy — good enough" style.

As I walked up the works drive on this historic day I reflected that this was to be the last one of thousands and thousands of times that I had proceeded along this drive to start work. In rain, fog, snow, black-out, and air-raids I had walked and even run along here to be at my job on time. I reached my office, and signed on in the register as usual, well in time, and at 8.30 prompt I started work as usual in a final endeavour to leave all the records, drawings, and jobs under my control but not yet completed in a tidy fashion and in a fit manner for my successor to continue with.

At 10.30 a.m. I was requested to go to the Chief Draughtsman's office, where, on arrival, I met my own section chief as well as the Chief Draughtsman. After a cordial greeting with these two gentlemen, we all proceeded to the office of the Mechanical Engineer (Development). This gentleman, who was in charge of the department in which I had served, greeted me with enthusiasm, and began conversing about my long career on bus work and the many changes I had seen take place. It is not usual for an employee to have the opportunity to tell the "boss" about very many changes and events that had taken place in the undertaking in the past and of which he was quite unaware. He was naturally very interested. After we had been conversing for some time, coffee and biscuits were served, and after these had been partaken of he presented me with a sealed envelope. This contained the sum of money which had previously been collected from the staff in the drawing and technical offices. He also gave me a retirement card signed by all my colleagues in these two offices.

I had previously expressed my wish to the Chief Draughtsman that I did not want an office presentation, and that any money collected on my behalf should come only from the technical staff in these offices. This wish was graciously conceded. The reader, after reading my views expressed in the section entitled "Retirement from Bus Work", can probably understand why. I now explained to my chief that with the money which was given to me I intended to buy a tea trolley, and this I did indeed purchase the next week. After the presentation had been made I was thanked by my chief on behalf of the Board for all my long years of service. He expressed a wish that I may enjoy a long and happy retirement, and this was heartily endorsed by the other gentlemen present. After a final word and a handshake I said goodbye to all. The whole meeting pleased and satisfied me, and I felt the occasion was sincere and yet not over-done so as to cause me any embarrassment.

The only disadvantage of not having an office presentation that I could see was that I had no opportunity of publicly thanking all those who had contributed to my gift. I rectified this later, however, by writing a letter of thanks addressed to the chief to be posted on the office notice-board. There was a long-standing tradition in our office that on the occasion of a wedding or a retirement the person involved must pay for tea and cakes for all the staff from the chief to the messengers. I am sure this was always very much appreciated by all concerned, and I was naturally pleased to make arrangements for this tradition to be pursued on this occasion of my retirement.

After the meeting with the chief I started to move around the office saying good-bye to all my colleagues. This is not always such a happy event as many would wish us to believe, especially as I had known most of them for very many years. I have seen men's eyes filled with tears at farewell presentations, and when saying good-bye to colleagues. To some people the emotional strain is very great, and all too apparent. In my own case I was conscious that I had enjoyed a good innings of working life, for which I was truly thankful, but like all aspects of life it must come to an end. I had enjoyed my work in the office, and the atmosphere had been very friendly and pleasant.

After my tour of the office, the print room, and the stores, I proceeded into the factory to bid farewell to all my friends and associates there. Here also, as in the offices, many people expressed the wish that it were they who were retiring, and I wondered what their thoughts and emotions would be when it came to their turn to say good-bye to all the various problems associated with their lives in the numerous spheres of Bus Work on which they were engaged. Meanwhile my day sped on, and finishing time was fast approaching. I had not been able to see all the people whom I had hoped to, but at least I had done my best, and I felt reasonably satisfied. In due course I left for home soon after finishing time, for the very last time, and so ended the last day of my career in Bus Work that had extended over half a century.

Service certified

Some weeks after retiring I received a letter from the staff office asking me if I would be prepared to go to 55 Broadway, Westminster, to have an audience with the Chairman of the Board, and to receive my service certificate from him. I answered, and accepted the invitation, and on the appointed day I arrived at the staff office at 3 p.m. Here I met five other retired veterans from the Board, and one of them was a fellow whom I had first met at Willesden Garage as long ago as 1920. All of them had completed over 50 years of service with the undertaking. We were all greeted by a member of the staff office, one by one, and then escorted in a group to the Chairman's office situated on the top floor of the building.

It appeared to me that this place was regarded by so many people working in the building, as "the Holy of Holies", into which to gain entrance was an achievement worthy of special note. However, we were introduced to the Chairman, and we found him extremely nice and quite informal. We all sat down to a quiet chat and a general discussion embracing the past experiences of all the various people present. Tea and biscuits were served, after which our service certificates were presented to us, and then with a parting handshake from the Chairman we all bade him farewell. I thought it was very good for such an important man as the Chairman to give us some of his most valuable time. I liked the whole procedure, for it pleased me, and I thought it was a very fitting way in which to end one's career with such an organisation as London Transport.

The certificate, in an unpolished oak frame, measures about 12 in. x 18 in., and is of colourful design. It incorporates with oak leaves and scrolls the famous London Transport "Bulls-Eye" sign and Griffins, together with the county arms of the counties in which London Transport services operate. The whole design is very pleasing to view, and I think the Board is to be congratulated in adopting it. The wording can be seen overleaf.

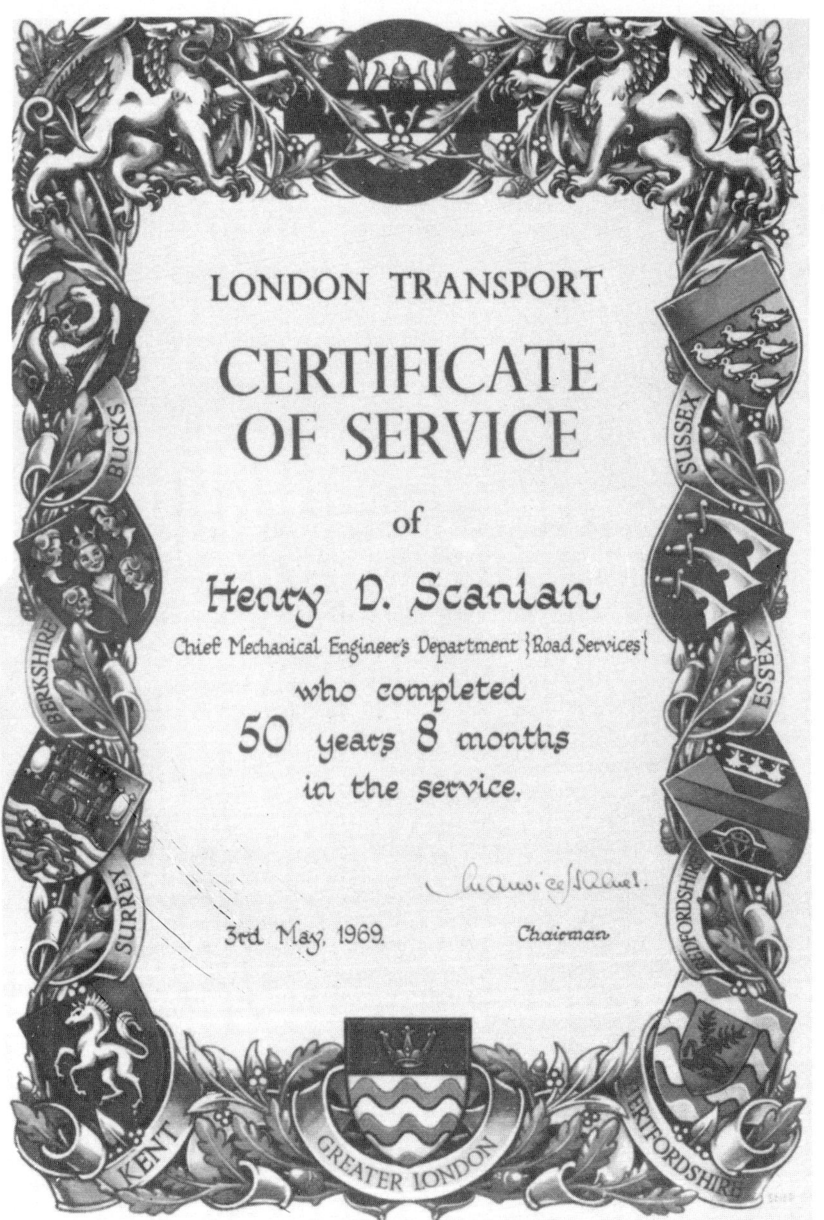

I am not going to satisfy the reader as to what I have done with my certificate, but it is certainly not as I have heard whispered at other presentations I have attended in the past, "Hang it in the smallest room". I know of one veteran of the Board living not far from me, who has his certificate hanging in the hall of his house for everyone to see, and even casual callers at the door cannot fail to see it, for he is justifiably very proud of it and of what it signifies.

Conclusion

In ending this narrative, I should like to express my grateful and sincere thanks to my wife for all her work and effort in typing the greater part of my original manuscript. To my friends and colleagues on the technical staff, Mr. A.R. PRITCHARD and Mr. C.J. PADLEY, I take this opportunity to tender my sincere thanks for reading my writings, and for their continued encouragement, with most valuable comments and criticism. I should also like to express my appreciation to my employing officer in London Transport latterly, Mr. A.R. PURVES (the Mechanical Engineer — Development, Road Services), for reading this book and for his most valuable assistance and suggestions so spontaneously given. To a very old friend of mine, JOHN C. GILLHAM, whom I have known for well over 30 years and who formerly worked with me at Chiswick, I herewith record my most grateful thanks and appreciation to him for editing and retyping the entire work, and for all his assistance to me in preparing this story ready to submit to my publishers. I am certain that without his valuable help and experience I could not have achieved my ambition to complete this work, at least not in its present form.

To you, my readers, I trust that you have enjoyed reading this authentic record of half-a-century of "bus work", and have at least derived some little knowledge from so doing. To the young man entering bus work as a career today, I would say, study and work hard, and gain all the qualifications and experience that you possibly can. These cannot be bought, and when gained they cannot be lost or taken away. Be tolerant to the Chief and fellow-workers alike, thus commanding the respect of all. And above all, lose no opportunity for advancement if you so desire.

SCANLAN

Produced above is the monogram used by the author on thousands of drawings, sketches, schemes etc., that he has prepared during his career in the Drawing Office.

This has been deliberately done first and foremost to let all interested parties know from whence the work emanated, and to save valuable time should any queries arise.

Secondly in an endeavour to produce such work that he could be justly proud to be associated with and marked with this monogram as a 'Hall Mark' of the highest quality or standard of work that time and circumstances would permit.

To the 'would be' Draughtsman, Technician, or 'Penpusher' in the realm of 'Bus Work' never be scared of signing a job well done. Do not hide your identity by a silly or absurd scrawl. Be mindful that persons who unfortunately cannot read or write sign their names with a cross.

Above all be proud of any task you have well and truly accomplished.